SUPERHERO PHYSIOLOGY

Kevin Boldt, PhD

Superhero Physiology
Copyright © 2021 by Kevin Boldt

Tellwell Talent
www.tellwell.ca

ISBN
978-0-2288-5549-1 (Hardcover)
978-0-2288-5548-4 (Paperback)
978-0-2288-5550-7 (eBook)

Science is full of heroes. Isaac Newton famously said that he was able to see further only by standing on the shoulders of giants. There are giants in science who have changed the way we understand the world around us and how we think about ourselves. Some of those scientists are discussed in this book, but science is also filled with many personal heroes of mine. Though none of them have superpowers (at least that I know of), I have countless heroes who have mentored me, invested in me, and inspired me to pursue knowledge, while teaching me how to think critically, how to design and conduct experiments, and how to learn from failure. There have been far too many to mention, but **this book is dedicated to my heroes.**

TABLE OF CONTENTS

FOREWORD

Superheroes have had a long relationship with science. Many of our favourite characters have received their abilities as the result of a medical procedure, genetic mutation, scientific accident, exposure to some radiation or bioagent, or their abilities come from a feat of engineering. This book is my attempt to consider how several of these abilities could be achieved for our heroes, considering our current scientific understanding of human physiology.

Much of my own research has been targeted at understanding how our bodies adapt to exercise and/or dietary intake, and how these adaptations can be enhanced to allow us, as humans, to either be healthier or to perform at a higher level. Given that many fictional superheroes are humans (at least, humanoids) with enhanced abilities, in many ways they serve as the ultimate examples of pushing the human potential to its limit. As a researcher interested in enhancing human physical performance, I cannot help but have an interest in superheroes as a model for exploring extreme human potential.

When I watched *Captain America: The First Avenger*[1] for the first time, I was intrigued by the concept of a super serum, and the incredible effects it had to enhance Steve Rogers' muscle mass, physical strength, jump height, and running speed. I found this transformation fascinating because these effects are many of the same factors that I seek to help the athletes I work with improve through training, and while these factors adapt with hard work over time, Steve Rogers turned into Captain America in one minute and 55 seconds. I was entirely distracted through the second half of the movie because I was engaged in thinking about how exactly the serum could trigger those changes and so quickly. I considered what acute adaptations to the tissue compositions, body structures, or metabolic processes could be changed that would lead to such sudden and incredible alterations in function.

In science, we are constantly trying to determine the mechanisms of action. It is not enough for us to know *that* something works; we want to know *how* it works, either because we are simply curious or so that we can optimize the effects. A real world example of this would be muscle growth in response to resistance training. At least as far back as ancient Greece, humans have been training professionally for sport. The resemblance between classical Greek sculptures and modern athletes as evidence of muscular development suggests that the ancient Olympians were utilizing resistance-based training to develop musculature and improve their performance at the games. We now understand that muscle growth is stimulated by both the accumulation of metabolic by-products (cue the burning sensation you feel in your legs when running up a set of stairs) and by mechanical stress detected by specialized fibres within the muscle (when under a heavy load such as lifting heavy weights).[2,3] Since we understand that muscle growth can be triggered by each of these two factors, we typically divide weight training programs into periods emphasizing metabolite accumulation (low weight/high repetitions), mechanical stress (high weight/low repetitions), or a combination of both (moderate weight/repetitions). In this way, we use our knowledge of how muscle growth is stimulated to enhance its overall adaptation, giving us an advantage over our ancient Greek ancestors. To return to our super serum, I think most people were content to accept that a super serum could turn Steve Rogers into Captain America simply because of science. However, scientists like myself are dissatisfied with "because of science" explanations. Since my initial viewing of Steve Rogers' transformation, I have greatly considered the theoretical physiology underlying how these, and other, super abilities could be achieved. I will share these considerations with you in this book.

Before I go any further, I'd like to preface that I am a believer and not a pessimist. You will not read here a lengthy explanation of how superheroes are unrealistic or how any particular superpower would not be feasible. Instead, I will propose ways an ability could theoretically be achieved. I do concede the caveat that for the sake of enjoyment, I do, in some cases, simplify (maybe over-simplify) a few scientific principles when I propose mechanisms for these super abilities.

The basic structure for each chapter begins with a description of the enhanced physical ability and the parallel in human physiology. Then, I will propose mechanisms whereby human physiology could be altered in ways that would result in the enhanced super ability, followed by additional considerations.

The focus of this book is on human physiology – the biological functions of the body. However, in many cases, as is the nature of science, the focus overlaps other disciplines such as physics and chemistry to help explain many of the mechanisms and implications of these abilities. Further, I have tried to avoid the use of math as much as possible, but since there are times where it's quite unavoidable, I have attempted to at least keep concepts and details as simple as possible.

My graduate supervisor was fond of saying that if you truly understand something then you should be able to explain it simply enough for your grandmother to understand (no offence, Grandmas!). Therefore, I endeavor to tread the line between writing an accurate description of human physiology with sufficient detail to stay true to proposed mechanisms for each ability and writing something that can be understood generally. I suppose it will ultimately be up to you, the reader, to determine whether I have been able to stay on the right side of that line, but I hope you will enjoy this book.

CHAPTER 1

Super Strength

Arguably the most common ability of superheroes is that of enhanced strength. How else could our hero stop a speeding train, lift a bus, or break through walls? The utility of superhuman strength can hardly be understated and is often what turns the tides when our heroes are pressed to their limits.

We must begin a discussion of super strength with a description of what we mean by strength. All animal movement, humans included, is generated by muscles. Simply, when a muscle is activated and produces force, the muscle pulls on its tendon (tendons connect muscles to bones), and the tendon transmits the muscle's force to a bone, which rotates or tries to rotate. For example, if I activate my biceps muscle, it pulls on my biceps tendon, which pulls on my forearm, and my elbow flexes (bends). If the limb that I am trying to move is restrained by something, like a weight, then my muscle needs to produce more force to move the limb. To continue our example, if I grab my cup of coffee, my biceps muscle will need to produce more force to overcome the load and allow me to bring my cup to my mouth for a drink. Further, if I grab a dumbbell, my biceps will need to produce even more force to move the load. If I continue to increase the load by grabbing even heavier weights, there would be a point where the weight of the load exceeds the ability of my biceps to produce sufficient force to move it. Maximum strength is generally defined as the maximum force that an individual can produce for a given joint. The elbow in this example. Therefore, when we talk about super strength, we are referring to the ability of the individual to produce superhuman muscle forces. Before we address how super-level forces could be produced, we need to talk about how muscles generate force in the first place. In order to understand force generation, we need to take a look at the structure of muscle.

Muscle Structure

Muscular force is generated from deep within the muscle. Muscles are beautifully structured in a distinct hierarchy that is comprised of a series of bundles. If you follow along with Figure 1.1, a whole muscle is composed of bundles called fascicles; fascicles are composed of bundles called fibres; fibres are bundles composed of myofibrils; and myofibrils are long chains of sarcomeres arranged end to end. It is this serial (end to end) arrangement that gives the myofibril its striped pattern.

Fibres are the name we give muscle cells and are the basic physiological unit of muscle. These specialized cells are very thin (approximately 1/5 of a mm) but can be quite long (up to 13 cm). Since they are cells, they contain all the organelles you learned about in high school biology: nucleus, mitochondria, ribosomes, etc. Although, because muscle fibres are so long and they undergo constant repair, they contain several cell nuclei. Additionally, since they require a substantial amount of energy to sustain their contractions, they have a much greater density of mitochondria (the powerhouse of the cell!) than do other types of cells. The defining factor of what makes a cell a muscle cell is the bundles of filaments running the entire length which we call myofibrils. Myofibrils are very tiny and, when combined in bundles, are the contractile force generating components of the muscle. Myofibrils are long chains of sarcomeres attached to each other end to end. At last, we have reached the sarcomere, which is the basic functional unit of muscle and is where all the magic happens.

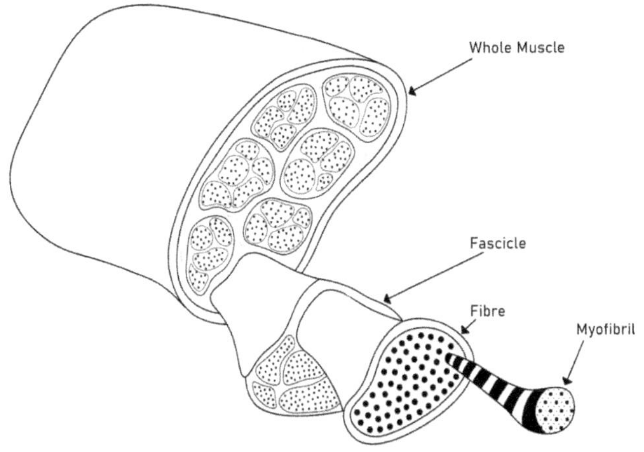

Figure 1.1: Muscle structural hierarchy from whole muscle down to myofibrils.

The Sarcomere

Sarcomeres contain the mechanical machinery powering your muscles and allowing you to move. Sarcomeres are comprised of two primary types of overlapping filaments: the thick and the thin filaments (Figure 1.2). The central thick filaments run lengthwise along the centre of the sarcomere and are surrounded by thin filaments. The ends of each sarcomere connect the thin filaments between adjacent sarcomeres that form into long chains (Figure 1.2 shows a simple chain of sarcomeres). The thick filament is able to interact with the thin filament via projections called crossbridges that branch off of the thick filament (which we will come back to shortly). The thick filaments are secured in place by molecular springs that connect each end of the thick filament to either side of the sarcomere. This spring prevents the sarcomeres from being pulled apart and keeps the thick filament centered between the thin filaments. The shortening and lengthening of muscle is referred to as the sliding filament theory because the length of the thick and thin filaments themselves don't change, but sarcomeres shorten or elongate as the filaments slide along each other (Figure 1.2). The extension of a sarcomere is achieved by the sliding of the thin filaments along the thick filament. Conversely, when shortening occurs, the thin filaments slide to the centre of the sarcomere and the whole unit shortens. Muscles are comprised of chains of thousands of sarcomeres, but Figure 1.2 shows an example of a simple, three-sarcomere-long muscle shortening and elongating by the sliding of the filaments along each other.

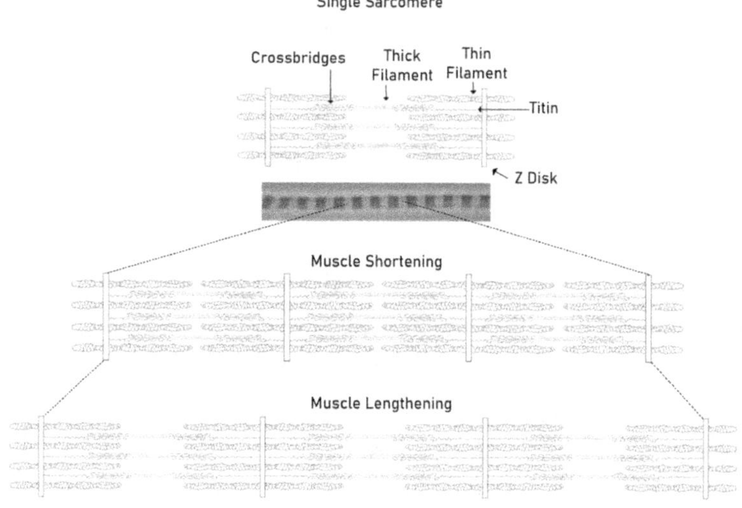

Figure 1.2: Top: Overlapping thick and thin filaments make up the sarcomere structure. Bottom: a simple chain of sarcomeres demonstrates how overlapping of thick and thin filaments allow for its shortening or elongation by sliding along each other.

Muscular Force

Our understanding of muscle function is due largely to the work of two giants in the field of human physiology. In 1954, Andrew Huxley and Hugh Huxley (no relation) published papers in the same journal describing their experiments on muscle where they independently came to the same conclusion.[4,5] Andrew Huxley (along with Rolf Neidergerke) did his experiments with frog fibres, while Hugh Huxley (along with Jean Hanson) did his using frog myofibrils. When they viewed the muscle preparations through microscopes, they noted distinct alternating dark and light bands, the same dark and light bands you can see in the microscope image of Figure 1.2. When the samples were stretched, the dark bands remained the same size, but the light bands increased in length. In their papers, they both concluded that the dark regions were the thick filaments (thick filaments look dark because their thickness blocked light), and the light bands were the regions with only thin filaments (light could pass through). As the muscle was stretched, the thin filaments slid along the thick filaments. Thereby, the dark region (thick filaments) did not change length, while the light bands increased as the overlap of the two filaments decreased. The experiments were simple, but the implications were profound. Two groups led by individuals with the same surname, doing almost the same experiment, came to the same conclusion and published their findings in the same journal at the same time, and the sliding filament theory was born. Over the next thirty years, this work was continued and developed into our current understanding of muscle function. Working off the basis that when muscle is activated the filaments slide along each other to allow the muscle to shorten, it was discovered that crossbridges were responsible for the force production of muscle.

Crossbridges, remember, are the projections that reach off the thick filament to interact with the thin filament. The crossbridge has two distinct sections: the long arm region and the pair of heads. Along the thin filaments are sites that have a strong affinity to the crossbridge heads and when these sites are exposed, crossbridges will bind to them. In the resting state, these binding sites are covered up with a regulatory protein, and as a result, the crossbridges are unable to bind to them and the muscle stays relaxed. When an individual decides to move a muscle, a signal is sent from the brain, down through the spinal cord, along a nerve, to where the signal reaches the muscle (more on this process in Chapter 3). When this signal reaches the muscle, the muscle responds by flooding the insides of its cells with calcium. Calcium-binding pulls the regulatory protein away and exposes the binding site. Once the crossbridge binds, it uses up a molecule of adenosine triphosphate (ATP; i.e. energy) and the crossbridge head rotates (Figure 1.3).

ATP consists of a main adenosine molecule and a chain of three phosphate molecules. The energy for a single crossbridge action comes from breaking down a molecule of adenosine triphosphate (tri = three) into adenosine diphosphate (di = two). The removal of one of the phosphates liberates energy stored in the chemical bond, and the crossbridge uses that energy to rotate. This rotation pulls the thin filament so it slides along the thick filament. The crossbridge head then releases, is fed a new ATP molecule, and if calcium is still present, the crossbridge binds again to the thin filament. As long as the muscle is active and energy (ATP) is available, the heads all along the thick filament continue to cycle and, stroke by stroke, pull the thin filaments along the thick filaments, sliding the filaments along each other, producing force, and shortening the muscle. To simplify, let's imagine this process in terms of a Venetian gondolier. Throughout the canals of Venice, gondoliers navigate their boats using a long pole. To move forward, they stick their pole into the ground at the bottom of the canal, leverage it, and propel their boat forward. By repeating this process, they can navigate their boats, push by push, through the dizzying maze of canals. Correlated to muscle, if we imagine the pole is the crossbridge, the bottom of the canal is the thin filament, and the boat is the thick filament, the process is fairly similar with the exception that in muscle the thick filament remains stationary and the thin filament slides over it.

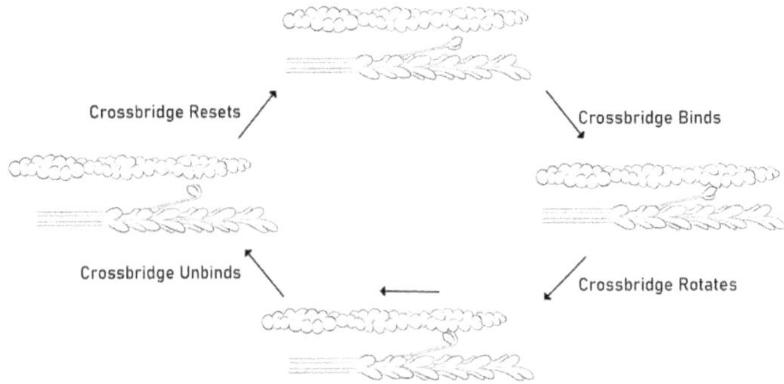

Figure 1.3: Schematic of the crossbridge cycle. Binding, rotating, releasing (unbinding), and resetting for the next cycle.

In some careful experiments, as they are very, very small, crossbridges have been isolated to reveal some meaningful results. A scientist named Jeffrey Finer and some of his colleagues broke apart a muscle and isolated single crossbridges in a petri dish, and allowed them to bind to suspended fragments of thin filaments.[6] They allowed the two molecules to interact and measured how much force was produced when the single crossbridge rotated. They determined that crossbridges consistently produce, on average, 3.4 pN (piconewton) of force. Newtons (N) are units for force and, for example, a 100 g mass exerts a force due to gravity (weight) of ~1 N. A picoNewton is 10^{-12} or 0.00000000001 N. What this means is that in order for a muscle to lift a weight, you would need a lot of crossbridges producing force! What this also means is, since the force each crossbridge produces is finite, the amount of force a muscle can produce is dictated by how many crossbridges it has active. Simply put, the total amount of force produced is 3.4 pN multiplied by the number of active crossbridges. Gains in strength resulting from resistance training are due to increases in muscle size. The muscle responds to resistance training by increasing the number of myofibrils within its fibres, and, therefore, the number of crossbridges that contribute to force. This is why larger muscles can produce greater force.

Super Strength

One thing comic book artists have in common with ancient sculptors is that they love to capture "idealized" human physique. Figures in Greek and Roman sculptures and comic book heroes all tend to have big muscles. The developed musculature of these heroes would make them very strong. Where our heroes surpass our understanding of human physiology is that superhero strength is usually disproportionate to muscle size. I will concede that many of our heroes are depicted with very impressive musculature. However, though actors like Henry Cavill, Gal Gadot, or Chris Hemsworth also have impressive physiques, they still need special effects to allow them to perform the feats of the characters they portray. The very strongest humans pull planes in strongman competitions if they pull with all their might, but this is nothing compared to the abilities of our heroes to effortlessly lift them. Human muscular force is largely proportional to muscular size. However, superhero strength is not. The following sections will consider how superhero muscles could disproportionately produce greater force than should be able.

Increased Duty Ratio

If you watch elite rowers compete, every rower in the boat pulls their oar at the same time. In perfect synchrony, each rower dips their oar into the water, pulls it back, and the boat propels forward. This works very well for rowing since as the boat is able to glide along the water, it carries momentum between strokes. However, in situations where momentum does not carry, such as if you were to do the same thing when walking, moving both legs at the same time, you would be doing two-foot hopping, and you'd lurch forward instead of walking smoothly. If you now think about what I described for muscle, if all the crossbridges in a muscle were to cycle (bind, rotate, and unbind) synchronously, the result would be jerky as the muscle would shorten in discrete movements (as if the crossbridges were "hopping" along the thin filament). This would result in inconsistent, jerky, hard-to-control movements. If you watch someone in a gym struggling to lift a weight at the end of a set, you may notice that their arms begin to jiggle, and movements become shaky. This trembling is the result of the crossbridges synchronizing to maximize force output in the onset of fatigue. Though synchronized action works optimally in rowing, it does not work well for crossbridges within a muscle because we rely on smooth coordinated movement, and momentum does not carry between crossbridge strokes.

To compensate for this, crossbridges cycle out of phase with each other. By doing so, some crossbridges are bound and producing force while, concurrently, others are unbound, and by the time the first set unbinds, the other set has bound and is contributing to force. This results in smooth contractions. The downside to this is that, since we know the force of a muscle is proportional to the number of crossbridges bound at a given time, desynchronizing action will limit how much total force can be produced.

The proportion of crossbridges that are bound is referred to as the duty ratio, where a higher duty ratio reflects a greater proportion of bound crossbridges, and a lower duty ratio reflects a lower proportion. In fully activated mammalian muscle, the duty ratio is around 30%.[6] As I mentioned, this results in smooth muscle contractions, but it also means that 70% of crossbridges are somewhere in the detached part of the crossbridge cycle. Thus, our muscles only produce 30% as much force as they are capable of. If the duty ratio could be increased, a muscle would produce more force. This could never reach 100% because it would affect coordination and movement would be jerky, but if our hero could tune their muscle to increase the duty ratio (>30%), their force output could certainly be enhanced.

Crossbridge Force

If total muscle force is dictated by the number of crossbridges bound and the amount of force per crossbridge, then mathematically, total muscle force for our heroes could be increased by increasing either the total number of crossbridges or the force output of each. Crossbridge force is consistent around 3.4 pN, and if this force could be increased, it would have significant implications. With the incredibly high number of crossbridges that contribute to force production, an increase in force per crossbridge multiplies very quickly.

As mentioned previously, the energy for a single crossbridge action comes from breaking down a molecule of ATP into ADP. The removal of one of the phosphates liberates energy stored in the chemical bond, and the crossbridge recycles that energy to produce force. However, if each crossbridge were able to break two phosphates off ATP, turning it into adenosine monophosphate (AMP) (mono = one), more energy would be available for the crossbridge to use. Additionally, each crossbridge has two heads but only one of the heads normally interacts with the thin filament during a cycle. If both heads were able to bind and each crossbridge were to split an ATP molecule into ADP, or, better yet, into AMP, even more energy would be available. More energy means more force, and total force output could be substantially enhanced without changing the number of active crossbridges or impacting the fluidity of movement. More force per crossbridge means stronger muscles, and stronger muscles mean more villains get put where they belong!

Longer Filaments

A scientist named Albert Gordon, while working with friends (Andrew Huxley included), determined that the amount of overlap between the thick and thin filaments in a sarcomere dictates its force output.[7] At long sarcomere lengths when the filaments have little overlap, only crossbridges at the end of the thick filament can interact with the thin, so force output is low. At shorter lengths, the overlap is greater so more crossbridges can interact and contribute. At the optimal length, all crossbridges along the thick filament interact with the thin filament and the force output of the sarcomere is at its maximum. Crossbridges are 14.3 nm (1 nm is 0.000001 mm) apart along the thick filament. If the crossbridges were closer together (greater crossbridge density), then more thick-thin filament interactions would be possible for the same amount of overlap. However, if the crossbridges were too close together they could interfere with each other as they rotate and cycle. Rather, instead of increasing the density (decreasing distance between crossbridges), longer thick and thin filaments would allow for a greater number of crossbridges without overcrowding. If the filaments were to be longer, the potential number of interactions and force output would be higher. Simply put, if our hero had longer thick and thin filaments in their muscles, their maximum potential number of active crossbridges, and thus force output per sarcomere would be enhanced.

Activation of Titin

Along with the contractile elements of the sarcomere, the thick and thin filaments, there is the passive spring-like protein called titin (Figure 1.2). There are some interesting properties of titin still being discovered. For example, it was thought that titin acted as a purely passive spring to hold together the sarcomere and keep the thick filament centered. Recently, it has been demonstrated that titin can modulate its stiffness under the right conditions.[8]

If you activate a muscle fully, and then actively stretch the muscle (eccentric contraction), the force output of the muscle is actually higher than the maximal output a muscle can produce isometrically (with no change in length). This is aptly termed force enhancement, and one of the leading theories of how this occurs is through titin modulating its stiffness. The force carried by a spring is the product of how stretched the spring is and how stiff it is. If titin could change its stiffness substantially, then as long as it was partly stretched, it would increase the total force of a sarcomere. Therefore, the final mechanism proposed here is that titin would increase its stiffness, and this seemingly passive element would contribute to force production along with the crossbridges, making our hero even stronger.

Overall Force Enhancements

Each mechanism I have proposed could independently increase the force output of a muscle. However, the combination of these factors would be the most effective. For example, if our hero could increase the duty ratio of a muscle, produce more force per crossbridge, had longer myofilaments, and even a modest contribution of titin, their strength would be increased significantly in disproportion to muscle size, as is seen in our super-strong heroes.

Additional Considerations

Energy

Muscle activation requires energy (more on this in Chapter 9). The heavier the load that needs to be lifted, the greater amount of energy that will be utilized to perform that action. This has two implications. First, our hero would need the physiological infrastructure (overall physical fitness) to convert and deliver that energy to the muscles as ATP (again, in Chapter 9), and our hero would need to consume more food to supply the energy for ATP synthesis.

Force Scaling

The ability to scale the force output of a muscle is very important. Not to brag, but I possess the physical prowess to crush an egg with my bare hands. However, if I want to use that egg to make my breakfast instead of showing off, then I need to be able to carry the egg without crushing it. Exaggerate this ability to super strength and the potential to inadvertently damage property or hurt common folk becomes much more serious. Though our hero may have the strength for incredible feats, they need to be able to control it and use it in delicate situations such as rescuing innocent bystanders or making a morning omelette.

Soft Tissue Strength

It doesn't benefit anyone if our hero has the strength to punch through a wall or lift a plane above their head if their bones don't have the strength to withstand that kind of impact or to support the additional load. Super strength is meaningless unless our hero also has enhanced strength in their bones or their soft tissues like tendons, and cartilage. Muscles insert onto bones via tendons. When a muscle produces force, this force is transmitted through the tendon and pulls the bone. If a muscle produces a super-level force but the other tissues aren't also enhanced, then either the tendon will tear, the tendon will pull right off the bone, or the bone will bend and may break. At the very least, our heroes would certainly face ligament and tendon sprains and strains, or osteoarthritis as their cartilage becomes damaged or degraded from the high compression and shear loads. Since we don't see many superheroes with ligament sprains, arthritis, or broken knuckles from punching walls, we can assume their bones, tendons, ligaments, and cartilage have enhanced strength along with their muscles.

Conclusion

Super strength has the potential to solve a lot of our hero's problems. The most substantial difference between our hero and a normal human, is that the hero's strength is largely disproportionate to the size of their muscles. This could be accounted for by either increasing the number of active crossbridges within a muscle and/or by increasing the amount of force each crossbridge is able to produce. Add those up and you get incredibly strong muscles that endow our hero with the ability to stop a speeding train, catch a flipping bus, and break through walls.

CHAPTER 2

Super Speed

In 2008, Usain Bolt won the Olympic gold medal in the Men's 100 m sprint with a time of 9.69 seconds. Despite running for around 80 m and taking the final ~20 m of the run to celebrate, he broke the world record and became the fastest man alive. He went on to win several World and Olympic Championships (as well as a soccer career), ultimately setting the current world record of 9.58 s in 2009. If we do some quick math, we can determine that running 100 m in 9.58 s means an average running speed of 37.6 km/h. This is even more impressive in light that sprinters begin a race stationary and most of the 100 m race is spent accelerating. Researchers calculated Usain Bolt's maximum speed in these races to be close to 43.2 km/h![9] Truly an incredible feat. If Bolt and I were to run in a race (Boldt vs. Bolt; the Battle of the Bolts!), I wouldn't stand a chance. However, if we were to make Bolt compete against other members of the animal kingdom, he would fare about as well against them as I would against him. For example, he would not even be close to the podium if he was up against a greyhound (64.8 km/h), a horse (68.4 km/h), and a cheetah (104.4 km/h).[10,11] If you're like me, it is inconceivable to imagine running 100 m in less than 10 s like Bolt did. As a species, it is impossible for us to consider running at speeds over

100 km/h. However, when we imagine the speeds reached by our heroes, we imagine them travelling considerably faster than 40 or even 100 km/h. The question this chapter will address is how our heroes reach these super speeds.

Movement Speed

Movement speed is a function of stride frequency (how many steps you take in a given period of time) and stride length (how long each stride is). For example, if each step I take is one metre long and I take sixty steps in a minute, then I will be travelling 60 m/min or 3.6 km/h. If I wanted to travel 5 km/h then I would either need to increase my step frequency to 84 steps per minute, my stride length to 1.4 metres per step, or a combination of taking longer strides and faster steps. Super speed, therefore, could be achieved by either an increase in the length of each stride or the rate of steps.

Stride Length

Different animals use these strategies differently for locomoting at high speeds. A good example of utilizing stride length is the kangaroo. Kangaroos have fascinated researchers for many years, largely due to seminal work by Terence Dawson and Richard Taylor.[12] The pair observed kangaroos running/hopping on treadmills while measuring the animals' breathing to determine their energy consumption. As the treadmill was increased from very low speeds to pretty low speeds, the animals required more energy to maintain the faster pace. This, of course, is exactly as you would expect—more exertion is required to jog than to walk. However, as the animals continued to accelerate beyond 7 km/h, energy consumption did not continue to increase and, in fact, even decreased slightly! They discovered that it is just as easy for the kangaroo to hop at 7 km/h as it is to hop at 25 km/h. This has been attributed to the animal's long elastic tendons, which very efficiently store and return energy. As the kangaroos hop along, their tendons do all the work for them like a set of pogo sticks (more on this in Chapter 8). What is particularly interesting for us now is that stride frequency is the same whether the kangaroos are running 7 km/h or 25 km/h, so it is the length of their strides that accomplishes the greater speeds. On average, the kangaroos in Dawson and Taylor's study weighed 23 kg, and when allowed to roam freely, were clocked up to 65 km/h. Though this still isn't super speed, it shows that increasing stride length can be an effective mechanism on its own for increasing running speed.

Strength = Speed?

A dominant characteristic of a sprinter phenotype is the size of their lower limb musculature. Sprinters have big legs, which generate a lot of power. This power allows them to do two things very well: generate the propulsion to accelerate and to maintain stride length at high speeds. This high level of strength allows them to bound almost like a kangaroo as they propel themselves forward in long strides. If we consider the factors from the previous chapter, super strength alone could allow an individual to run at high speeds as they bound along with long strides. However, this would be limited as taking longer strides means more time between steps. Since steps are where individuals can change direction or adjust their trajectory, long strides would limit agility and the ability to change direction or slow down or speed up. When we expand this to the extreme, as we would need to do for super speed, the duration between footfalls would become too long to be effective. Our very strong hero wouldn't be able to change direction and would only be able to run fast in a single direction, like Dawson and Taylor's treadmill-bound kangaroos. Thus, super strength alone could account for a very high running speed as a result of long strides. However, given its limitations on maneuverability and reasonable limits of speed, super speed would have to rely upon rapid stride frequencies.

Mechanisms of Super Speed

The process of a single step during running involves the step phase and the flight phase.[13] The step phase is when a foot is planted, and the flight phase is when the foot is not in contact with the ground. The step phase can be further broken down into absorption and propulsion. Once the toe lifts off the ground, the flight phase has begun. At this point, the leg is behind the runner, and the ankle, knee, and hip are all extended. During the flight phase, the hip flexors and hamstrings activate to bend both the hip and knee. Even before the foot comes in contact with the ground, our muscles begin to prepare for the next landing: the thigh muscles (quadriceps) begin to extend the knee and leg muscles activate in preparation to absorb the impact of the landing. The heel strikes the ground and the lowerlimb muscles absorb the impact. As our runner's body passes over their planted foot, the quadriceps and glute muscles shorten and extend both the knee and hip, generating the forces required to propel the runner forward into the next step, completing the step cycle. The way to speed up this cycling is to increase the rate at which the muscles shorten, both during propulsion (producing force) and in the flight phase of pulling the leg forward for the next step. If muscles shorten faster, then the individual's legs will cycle faster. Thus, the root mechanism of increasing stride frequency for super speed is muscles that shorten faster.

In the previous chapter, I described crossbridge action within a muscle, which produces force by binding and rotating, and Jeffrey Finer's experiment of isolating single crossbridges and allowing them to interact with a suspended segment of thin filament.[6] Finer and his friends measured the average force per interaction (3.4 pN), but what I didn't mention is they also measured the average step length or how far each crossbridge pulled the thin filament per interaction. They determined that each crossbridge could shorten the sarcomere by around 11 nm each time it cycled.

Since each crossbridge produces 11 nm of shortening, the speed of muscle shortening is dictated by the rate at which crossbridges can cycle.

If we really break this down and we imagine a muscle has a single crossbridge, it can be likened to a canoe with a single rower. If our rower is already paddling as hard as they can, then the only way to speed up how fast the canoe travels is by increasing how fast our rower paddles. In our analogy, the canoe is like the thick filament and the water is like the thin filament. The only way for the filaments to slide along each other quicker is to increase the rate crossbridges can bind, rotate, unbind, and repeat. When we grow our analogy to a full muscle with thousands of crossbridges, the idea is the same. That is, the limiting factor for the speed of shortening is how quickly the crossbridges can unbind, reset, and rebind to produce another shortening. The factors responsible for providing the energy for crossbridge action are intramuscular proteins and enzymes. The primary enzyme, myosin ATPase, is responsible for breaking down ATP and liberating the energy to allow crossbridges to cycle. Thus, this primary enzyme for breaking down ATP (myosin ATPase) largely dictates the shortening velocity of muscles, by governing the rate crossbridges can cycle.

Myosin ATPase content varies between muscles and species, and to consider a practical example, let's compare the muscles of chickens and ducks. Muscle fibres can be divided broadly into either fast-twitch or slow-twitch fibres, and as you can imagine from the names, fast-twitch fibres have higher shortening velocities than slow-twitch fibres. All in, there are several subcategories of fibre types based on their contractile properties, but for simplicity, we will focus on the two main categories (described in the table below).

Table 2.1: Comparison of slow-twitch and fast-twitch muscle properties.

	Slow Twitch	Fast Twitch
Shortening velocity	Slow	Fast
Force	Low	High
Fatigability	Low	High
Blood supply	High	Low
Mitochondria density	High	Low
Myosin ATPase concentration	Low	High

When you're carving your roasted chicken and you place the breast next to the drumstick, you will undoubtedly notice something very different between the colour of the two cuts of meat. The dark meat (drumstick) is made predominantly of slow-twitch fibres, while white meat (breasts) is predominantly made up of fast-twitch fibres. However, in a duck, both the breasts and legs are dark meat. Myoglobin, a protein that supports oxidative energy metabolism, is abundant in slow muscle and limited in fast muscle and gives dark (slow) meat its red colour. A comparison of the lifestyles of a typical chicken and duck provides insight into why this is. If a chicken has to cross the road (I won't speculate why) she will most likely strut her way across. For this task, she isn't in a hurry, so she can meander her way across using the slow-twitch muscles of her legs. However, if a fox were to jump out and scare our chicken, she would flap her wings like crazy and fly a short distance to evade her predator. Chickens use the slow-twitch muscles of their legs for walking and the fast-twitch muscles in their wings for short flights where they need to flap their wings very rapidly to evade dangerous situations. Conversely, ducks use their wings to fly thousands of kilometres when they travel south for the winter and migrate back north for the summer. When they hang out in their favourite pond, they use their webbed feet to gently paddle themselves around. When the same fox scares them, they will either swim casually away or take flight and fly far away.

The duck locomotes with both its breasts/wings (flight) and legs (swimming and walking). These differences in locomotion and evasion strategies are the reason for the differences in muscles. Or, perhaps the differences in muscles are the reasons for the differences in evasion strategies (a chicken or an egg type situation).

Fast and slow fibre types are so named because slow-twitch muscles contract slower than fast-twitch muscles. Slow-twitch muscles contract at 1.2–1.3 FL/s (fibre lengths/second), while fast-twitch muscles are able to shorten at 3.6–4.4 FL/s.[14–16]. These numbers may not mean much to you but what is important is that fast-twitch muscles shorten at around 3-4 times the speed as slow-twitch muscles.

Despite shortening very fast and producing a high amount of force, fast-twitch muscles fatigue very quickly, while slow-twitch muscles are very resistant to fatigue. If a duck were to have the breasts of a chicken, they would become very fatigued, and they would hardly migrate 100 m, let alone 1000 km. Similarly, if chickens had the breasts of a duck, their wings would beat too slowly for a quick getaway and would end up as our fox's lunch. The reason slow-twitch muscles resist fatigue so well is because of their higher rates of blood supply, mitochondria density, and myoglobin content, all of which allow the muscles to be supplied with the energy required for sustained activity.

To return to the point, the real difference we are concerned with between slow and fast-twitch muscles is the myosin ATPase activity. With a higher shortening velocity, there is also a higher concentration of myosin ATPase as the ATPase allows for more rapid crossbridge cycling and greater muscle velocities. Let's dig further into examples from nature.

Super Speed in Nature

This chapter began with a comparison of animals able to run at high speeds. These running speeds are accomplished via fast-twitch locomotor muscles. Other animals have highly specialized (non-locomotor) muscles that can shorten even faster. For example, the rattle on a rattlesnake tail oscillates back and forth 90 times per second. Additionally, a very funny-looking fish called a toadfish makes a mating call that sounds like a vibrating cellphone by vibrating a membrane called their swimbladder up to 200 times per second! The rattlesnake's tail and the toadfish's swimbladder are controlled by highly specialized muscles that are designed solely for speed. A researcher named Lawrence Rome has done considerable work on these muscles that are aptly termed superfast muscles.[17,18] The toadfish's swimbladder muscle, for example, has a maximum shortening velocity of 12 FL/s, 4–12 times as fast as human muscle! Through careful experimentation on what makes these muscles unique, Rome and his friends have determined that in these muscles crossbridges can unbind from the thin filaments at rates up to ten times greater than other types of muscle. This allows the crossbridges to be reset, and to continue cycling at incredibly high rates. These rapid crossbridge cycling rates, and thus shortening velocity, can be attributed to high content of the primary enzyme, myosin ATPase, present in these muscles.

Nature's speediest runners, cheetahs, run up to 104 km/h[10]. Cheetahs have long bodies and limbs; they are light for their size; they have strong hind limbs, and a flexible spine, all of which enable them to take long powerful strides. They also have high proportions of fast-twitch muscle fibres to cycle their legs quickly. Running cheetahs have been measured to take strides up to six meters per step and up to almost four strides every second.[19] Thus, they accomplish these high speeds by utilizing both long strides and fast steps.

The Force-Velocity Relationship

Our heroes utilize their super speed to foil the plans of their villains. However, more villainous to our speedy heroes is the force-velocity relationship of muscle (Figure 2.1). First described in detail by Archibald Vivian Hill in 1938, the force-velocity relationship describes the inverse relationship between the force a muscle produces and the velocity at which it shortens.[20] For example, if you imagine trying to lift this book up into the air as fast as you can, you would likely be able to swing your arm very rapidly. Now if you were to try to lift the chair you're sitting on (assuming you're sitting) as fast as you can, your movement would be much slower, assuming you even can lift it. The reason this is relevant in this discussion is as our hero's legs cycle at the incredibly high rates necessary for super speed, the force output of their muscles would drop.

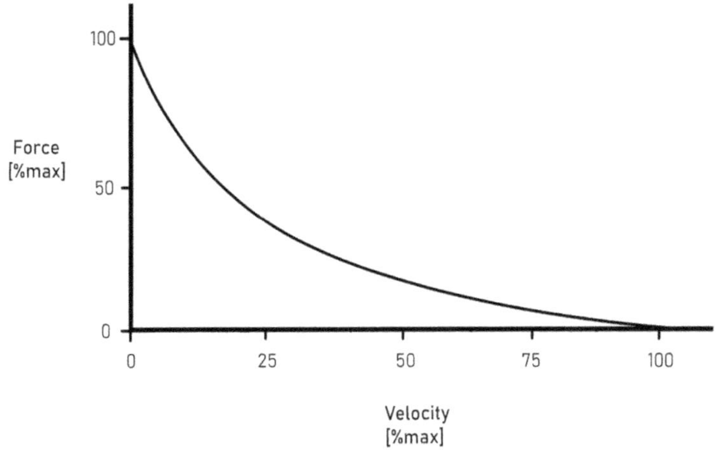

Figure 2.1: Schematic of the force-velocity relationship of muscle. The faster a muscle shortens, the less force it is able to produce.

The mechanisms for the force-velocity relationship are somewhat complex but can be simplified in terms of our cycling crossbridges. First off, crossbridge force is not instantaneous. During muscle shortening, the thin filament slides along the thick filament. If a crossbridge binds while the thin filament is already sliding past it, that crossbridge will have less time to contribute force to pulling. I'll use a tug-of-war as an example. Think about joining a tug-of-war where your legs can't move so to win you have to pull on the rope hand over hand. During a deadlock, your hands don't move, so you are able to pull your hardest. As your team begins winning and the rope slides past you faster, you grab the rope to get a good grip but the rope passes you and you have to release it for more rope further up. Your contribution at that higher speed becomes much less. If the rope accelerates, you would hardly be able to pull at all before you need to let go. If your team represents the thick filament, you represent the crossbridge, and the rope represents the thin filament. That is the reason for the lower force output of muscles at high speeds of shortening. At high shortening rates, crossbridges need to unbind before making their full contribution. This contribution decreases exponentially with increasing speed and results in the parabolic curve we observe.

What this means for our hero is that as they run faster and faster, and their muscles shorten faster, their muscles actually become weaker. At higher speeds, generating the propulsion necessary to maintain their step length is more challenging. Optimal power production of a muscle occurs around 30% of the muscle's maximum shortening velocity.[21] If our hero had muscles that shorten really fast, but they only had the strength of you or me (let's not over flatter ourselves), they would not be able to utilize the high shortening rates because they would be too weak to generate sufficient propulsion, especially when accelerating.

If you have ever run downhill, you understand that it can be difficult to stop running downhill when you decide you've had enough. This is because the faster you are travelling, the greater your momentum. To stop your momentum, you need to apply a force to resist it, and the larger the momentum, the larger the force required. When travelling at super speeds, it would require substantially more force to either change direction or to slow down or stop. With super speed alone, and without super strength our heroes would be unable to change direction, slow down, or stop. Imagine running down a hill as fast as you can, and then trying to stop or change direction suddenly. I'd say there's a very high chance that your momentum would cause you to end up head over heels (a very undignified position for a superhero). Therefore, to compensate for this, our hero's muscles would need an enhanced force output to produce the forces required for propulsion on acceleration and braking on deceleration. Even the toadfish encounters this problem as their swimbladder muscle is substantially weaker than other skeletal muscles.[18] For the fish, this is OK because this muscle only needs to vibrate a membrane, not produce high forces as in locomotion. Therefore, in addition to muscles that shorten at super speeds, our hero would need muscles that are also super strong. Super speed cannot exist without super (at least enhanced) strength.

Additional Considerations

Fitness and Food

Muscular contraction is hard. When you walk (or run) up a flight (or three) of stairs, you'll likely start to feel your legs beginning to burn. Next, your heart rate will increase, and you'll find yourself breathing deeper. Keep going and the burning sensation will increase, your heart rate will go even higher, and you will begin breathing faster. These are all signs that your body is trying to supply energy to your working muscles (more on this in Chapter 9). As your leg muscles contract to propel yourself up the stairs enzymes in your legs are converting sugar into ATP, lactate, and hydrogen (hydrogen accumulation = burning sensation in your legs). Your heart beats faster to increase blood flow and you breathe deeper and quicker to put more oxygen into your blood as it circulates to your legs. Once that oxygen reaches your legs, your mitochondria turn it into ATP. All that for a few, actually a lot of, molecules of ATP! The importance of this is that as I explained in the previous chapter, each single crossbridge cycle requires a single ATP molecule. If crossbridges cycle faster, they will also use ATP supplies faster. Two things are important to take from this: fast cycling of crossbridges requires high rates of ATP supply and high ATP supply requires lots of nutrients.

When you walk around, your lungs take in oxygen and transfer it to your blood. Your heart then pumps the blood to your legs, and your legs take that oxygen and mix it with sugar to produce ATP. When you start jogging, this happens faster, and when you run this happens even faster. In fact, the maximum speed you can sustain at a run is limited by the delivery of oxygen. This is termed VO_2max or the maximal rate of oxygen consumption. If you were able to deliver and utilize a greater amount of oxygen for ATP production, then you could both run faster and sustain the speed longer. Reaching super speeds would not only require muscles to shorten at incredibly high rates but also require lungs able to take in large amounts of oxygen, blood able to transport large amounts of oxygen, a heart able to pump that blood into circulation, and plenty of mitochondria to convert oxygen into ATP. The incredibly high demand of the super cycling muscles would need to be supported systemically with an adequate energy supply. In other words, our hero would need to be incredibly aerobically fit; otherwise, they would be unable to sustain super speed workloads. Additionally, oxygen is combined with macronutrients (sugar, primarily, but can also utilize fats and proteins) to produce ATP. This means our hero would need to consume a serious caloric intake to sustain this energy supply.

Reflexes

If our hero is moving like a speeding bullet, everything else will, relatively speaking, come at them like speeding bullets. Imagine our hero running along when suddenly an old lady strolls into their path or the villain they are chasing drops a banana peel. It doesn't do our hero much good to move their muscles incredibly fast if they don't have reflexes to avoid obstacles. Our hero would need to detect and process these visual stimuli (more on this in Chapter 4) and adjust their movement (more on this in Chapter 3). Without proportionally faster reflexes, our hero would effectively be running blind and would be unable to respond to the world around them as they move rapidly through it.

Wind

High-performance cyclists and super-speedy superheroes have both realized an issue that arises when they travel fast: it gets windy. At low speeds, air resistance makes relatively less of a difference. When you or I run, we don't notice any air resistance unless it happens to be a particularly windy day. However, with wind resistance at higher speeds (like in cycling), aerodynamics plays a more critical role. The air resistance that restricts a cyclist is proportional to the cube of speed (speed to the third power: $speed^3$) that cyclist is travelling.[22] In non-math terms, air resistance is the biggest factor restricting speed, and its contribution increases as speed does. Cyclists attempt to overcome this by wearing tight-fitting jerseys, cool helmets, making the front profile of their bike as skinny as possible, and by riding close behind each other. They also take advantage by leaning forward over their handlebars. By adopting this crouch position, cyclists decrease their front profile to improve efficiency by as much as 35%.[22] Thus, if our hero wants to run at super speeds with any economy of exertion, they will need to pick up an aerodynamic suit. Fortunately, spandex suits (like in cycling) come with the job.

Moving at high speeds means wind blowing in their eyes. Motorcyclists wear face shields, cyclists wear glasses, and cars, planes, and trains all have windshields. This is because wind blowing in your face can be a big distractor. When our hero is running at super speeds to foil the plans of their villain, they can't be dealing with chapped lips, dry eyes, or bugs hitting them in the face. In addition to their aerodynamic suit, they may need to consider investing in goggles and stocking up on bug wash.

Conclusion

Super strength alone could result in fast speeds but is unlikely to support super speeds. Most likely, super speed would be a result of increased rates of crossbridge cycling, allowing for rapid muscle shortening and thus greater stride frequency. To achieve this, these superfast muscles would also need to have super strength to sustain force development when accelerating, changing direction, and decelerating. Additionally, our heroes would need to eat a lot of food to sustain their energy demands, would need to be incredibly fit, and would need to look good in spandex.

CHAPTER 3

Telepathic Perception

Very few heroes (or villains) can detect or control the thoughts of others. Consequently, telepathy is an exciting ability because superheroes who do have this ability are often among the most powerful. For example, the "X" in X-Men refers to Charles Xavier ("Xavier's Men"), the telepath who uses his abilities to seek out and mentor the development of other heroes. My dog's name is Xavier, and I often wonder what he is thinking, but since neither he nor I possess his namesake's abilities, this discussion will remain theoretical.

Thinking about Thought

Any discussion about the ability to read thoughts must first consider what thoughts are and what it is that is actually being read. Philosophers have been very interested in thought, at least as far back as can be recorded. Buddha (480–400 BC) described thoughts as a subconscious flowing stream comprised of material form, emotions, external stimuli, desires, and awareness.[23] He theorized that through the practice of meditation, our thoughts can be focused and directed like opening and closing dams to control the flow of the stream. Plato (429–347 BC) surmised that thoughts were the true reality and that the physical world around us was just a model of a higher reality (imitations that he termed "forms").[24] Galen (129–200 AD) depicted what he deemed the rational soul, and contended that organs throughout the body were responsible for memory, thoughts, decision-making, and sensation, and were all integrated by the brain.[25] Descartes (1596–1650 AD) famously wrote "Cogito ergo sum," or "I think, therefore I am."[26] He reasoned that we only exist because we have conscious thought, and that who we truly are is simply the sum of all our thoughts. Since I am an exercise physiologist and not a philosopher, I will bow out here and stick strictly to the physical manifestations of thoughts, leaving the existential to philosophers. For the time being, we will focus on the chemistry and biology of thought.

Physiological Basis of Thought

Wherever you are sitting right now, you are reading words. That seems a silly thing to even mention to someone reading a book, but to read something, there has to be something to read. In this case, it is the words written on these pages; there is a language. So, when our hero reads an individual's mind, there has to be a language that they can perceive. That "language" is the neurochemistry of your brain. Regardless of the esoteric ideas of what thoughts are, at the very basic level, thoughts are the result of patterns of firing nerve cells that we call neurons.

Neurons

Neurons, like all cells, contain all the organelles you will remember from biology class in high school that are responsible for keeping them alive (nucleus, endoplasmic reticulum, ribosomes, etc.) (Figure 3.1). However, neurons are highly specialized cells that have some very unique features.

Similar to muscle cells, neurons are very long. There are three main segments of a neuron: the cell body, dendrites, and the axon.

The cell body (or soma) is where the organelles are found and is the epicentre for keeping the cell nourished, healthy, and is the site for new protein synthesis and regeneration of the cell. Extending away in several directions from the soma are root-like extensions called dendrites that branch off in protrusions called dendritic spines. Opposite the dendrites, extending away from the soma, is the long extension of the axon. The axon is the segment of the neuron that gives it its length and allows for its function. It is along the axon that nervous impulses travel, and our thoughts are transmitted.

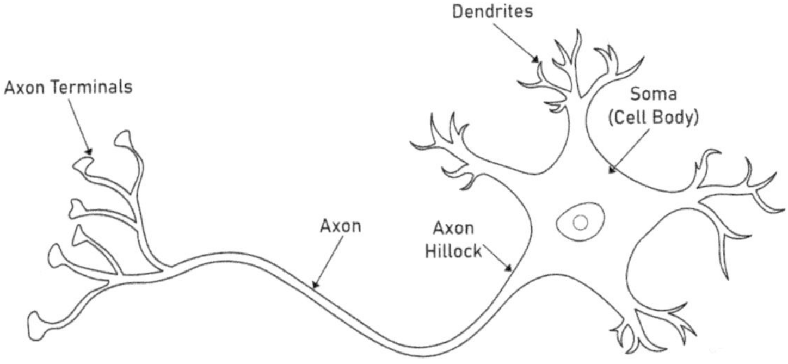

Figure 3.1: Schematic of a neuron's structure.

Nervous Impulses

Ions are the key to neuronal function. Ions are molecules that have an electrical charge, which can be either positive or negative. For example, chloride (Cl^-) has a negative charge, while sodium (Na^+) and potassium (K^+) have positive charges. In chemistry, these chemical charges are what facilitate ionic bonds. The salt you put on your eggs this morning was most likely in the form of sodium chloride, the sodium (positive) is able to bind to the chloride (negative) because of the attraction between their charges (more on this in Chapter 5). However, these charges have additional implications as well. Imagine a beaker (like the one on the left in

Figure 3.2) with a barrier down the middle that does not allow ions to pass through it. To the right of the barrier, we have plain distilled water, and on the left, we have water with dissolved sodium molecules. For simplicity, let's say we have ten sodium ions. When we add up the charges, there are ten "plus'" on the left (one for each Na^+) and zero on the right. If we were to measure the voltage of the distilled water (right), relative to the side with the ten sodium ions (left), we would read a voltage of -10 (arbitrary electrical units). The difference in charge between the two sides indicates its voltage gradient. If we were to add a gate (or channel) to the barrier in the middle, and we were to open it, the sodium would travel through it until it found equilibrium, five on one side, and five on the other – a neutral voltage gradient (like the right side of Figure 3.2). Let's apply this to nerve cells. If we stick an electrode inside of an axon and one outside of the neuron, the voltage measurement between the two tells us about the ion content inside and out. This is essentially the experiment done by Alan Hodgkin with neurons he got from a giant squid.[27] Squid have very large neurons (up to 1 mm in diameter!), so they were perfect for early experiments where equipment wasn't as precise as they are today. Hodgkin inserted electrodes into sections along the axon and recorded them. What he observed was an electrical pulse that moved along the length of the axon like a wave. His hypothesis, based on these observations, is the basis of our understanding of nervous function today.

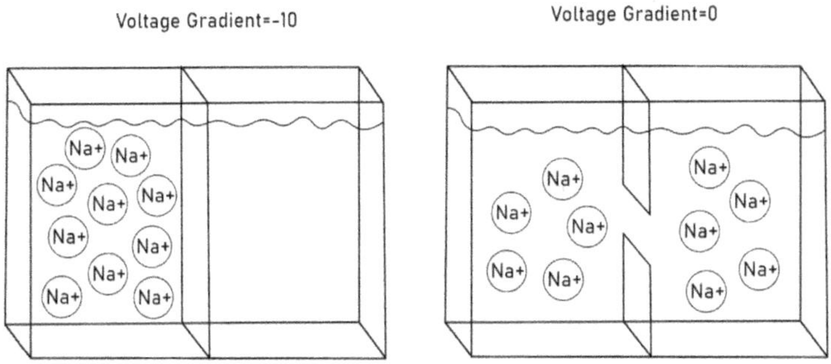

Figure 3.2: Depiction of a voltage gradient using a beaker with a non-permeable (left) and permeable (right) barrier. On the left, the barrier does not allow ions to transfer and results in a voltage gradient of -10 (the right side has ten less plus signs than the left side of the barrier). On the right, the permeable barrier allows ions to diffuse across resulting in a neutral voltage gradient (five plus signs on the left and five on the right).

What Hodgkin recorded was the transmission of an electrical signal conducted by the transmission of ions. At rest, sodium (positive) is highly concentrated outside of the cell membrane, and potassium (positive) is moderately concentrated inside. Since sodium is not able to diffuse through the membrane without an open channel, the inside of a neuron has a voltage of -70 mV compared to the outside. When a signal is detected by the dendrites from a neighbouring axon, it results in channels opening at the base of the axon (the axon hillock). This allows sodium to flow in and increases the interior voltage. When the voltage increases from -70 to -55 mV, an action potential is triggered. Such channels that allow ions to pass through are situated all along the cell membrane. We deem them voltage-gated channels because at rest they are closed but when stimulated by changes in voltage, they open. When the axon hillock reaches -55 mV, the channels nearby open and allow for an influx of sodium, increasing local voltage to +40 mV. Since only channels in proximity to the hillock open, it results only in a local change: the axon hillock is now +40 mV, while along the axon remains at -70 mV. However, this change in voltage triggers the next closest channel to open, which allows ions to flow in locally and that region depolarizes to +40 mV. Meanwhile, the sodium channels at the axon hillock close, and sodium is actively pumped back out of the region, decreasing the voltage back to -70 mV. The increase in voltage in the second region then triggers the next closest channel to open, triggering the next and the next and the next. In this way, the signal is conducted like a wave along the axon away from the soma (Figure 3.3). When this current reaches the end of the axon (axon terminal), it triggers the release of a neurotransmitter called acetylcholine, which is released from the first axon and travels to the dendrite of the next neuron. Once it reaches the dendrite of the neighbouring neuron it stimulates that neuron to fire and the signal is transmitted neuron to neuron. In this way, local changes in ion concentration propagate the signal down along the axon and eventually to the next neuron. Neurons are thus able to communicate to each other, allowing for signals to be sent between different parts of the brain.

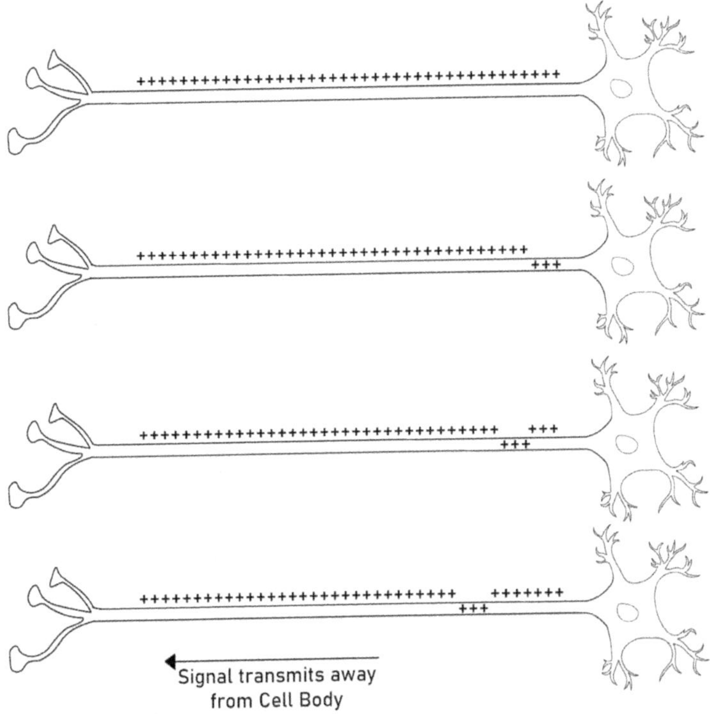

Figure 3.3: Wave of ion transfer across the nerve's membrane allows for transduction of the electrical impulse along the neuron. Sodium (+) is locally transported inside the axon, which triggers the opening of adjacent sodium channels. This wave of ion flow propagates along the axon away from the cell body.

The Brain

The nervous system can be divided into two predominant divisions: the peripheral and central nervous systems. The central nervous system contains the brain and spinal cord, while the peripheral system contains the sensory and motor nerves extending throughout the body. For the sake of telepathy, we will focus on the central nervous system and the brain. The brain is an incredibly complex organ comprised predominantly of neurons. The brain's distinct functions are fairly well compartmentalized, so it's easiest to address the brain's function by dividing it into its distinct regions.

Cerebral Cortex

The cerebral cortex is composed of valleys and ridges, which give it the characteristic texture that probably comes to mind when you think about the brain. The cortex is the most evolutionarily advanced portion of the brain and is responsible for high-level functions: thoughts, cognition, deliberate motor control, and decision-making. Structurally it is the most superficial part of the brain and makes up about 40% of the overall weight. There are two sides to the cortex: left and right. Throughout the cortex are specialized regions that are the centres for various senses and controls. For example, on the left side of the temporal lobe (side of your head above your ear), there is a region that processes language and is where your ability to speak comes from. Additionally, along the top of your brain is the sensory cortex, which interprets haptic sensation, and immediately in front of that is the motor cortex that generates the impulses that control movement.

Brainstem

The brainstem is the section where the brain meets the spinal cord. While the cortex is the most advanced part of the brain evolutionarily, the brainstem is the least. Due to its location, it is responsible for transmitting signals from the brain to the spinal cord, and vice versa. It also has specialized regions for subconsciously controlling breathing, heartbeat, and posture. Essentially, your brainstem keeps you alive without you having to think about it.

Cerebellum

Cerebellum is Latin for "little brain," and that's exactly what it looks like. The cerebellum, like the cortex, has both surface folds and two lateral hemispheres. The cerebellum controls postural adjustments and assists with repetitive locomotion. When you walk, for example, you don't need to consciously think about each step. Once you decide to walk across a room, you can without thinking, and that movement pattern is maintained by the cerebellum

Diencephalon

I mentioned previously that the cortex has two hemispheres, the left and right. The left side of your cortex controls and senses the right side of your body and the right cortex manages the left side of your body. If these hemispheres were allowed to function independently without communicating, you would be quite a mess. You would reach for things with both hands instead of just one, you wouldn't be able to coordinate walking, and making decisions would be an unending conflict. Fortunately, the two hemispheres are connected through the diencephalon. This structure is responsible for transmitting signals from the left hemisphere neurons to the right hemisphere neurons and back again. Deep within the diencephalon reside structures that control metabolic rate, body temperature, and hormonal balance.

The brain is marvelously complex: it stores your memories, allows you to move, dictates your personality, allows you to make decisions, and of course, it holds your thoughts.

The Nature of Thought

At the very basic level, all your thoughts, every part of your consciousness, all your memories, personality, and emotions are really just patterns of neurons firing across the regions of your brain. The patterns of neurons firing and how often, predominantly through and across your cortex, are your thoughts. Basically, who you are is just the transmission of sodium and potassium in and out of specialized cells. Talk about philosophical!

Sensation and Perception

When it comes to our senses, psychologists make the distinction between the two processes of sensation and perception. For our brain to process something (like how our hero detects the thoughts of the person next to them), we must first sense the stimuli, and then we need to interpret (perceive) what that stimulus means. Generally speaking, this is a two-step process. We will discuss this in more detail in Chapter 4 for sight (and the eye), but for the sake of the present chapter, we will use hearing (and the ear) as an example.

Sound is a series of vibrations that travel through the air. The reason sounds sound different from each other is a result of different wavelengths. Within our ears, we have several key structures enabling us to detect both the presence of and characteristics of the vibrations. If I clap my hands together, the collision between my palms creates a sound (vibration) that propagates through the air. Once those vibrations reach my ears, they enter my ear canal and terminate on my eardrum. The eardrum functions very much like a drum (hence the name), where the sound waves cause it to oscillate at the frequency of the sound waves. This vibrating membrane transmits the signal to three small bones (the smallest bones in your body), which then transmit the frequencies onto an organ deep inside your head called the cochlea. The cochlea looks like a snail shell and is basically a long tube wrapped up like a cinnamon bun filled with fluid. Lining this tube are hair-like cells that reach out like sea anemones. As the sound waves are transmitted from the three bones to this fluid, waves are propagated through the fluid at the same frequency as the waves that reached the eardrum. The hairs are sensitive to changes in pressure, so as the waves pass over the hair cells, they are sensed. The cochlea is tapered from the base to the tip and because it gets progressively thinner, signals with a low frequency (long wavelength) travel further along the tube than signals with a high frequency (short wavelength). The cochlear hairs detect the pressure signal and send a nervous signal to the brain that indicates they have detected sound. The detection of the sound stimuli is the process of sensation. The brain combines all the signals from the hair cells and, depending on which regions of the cochlea are reporting, the brain interprets these signals, and we perceive what we are hearing (perception). If a breakdown occurs anywhere along the line, then the sound cannot be "heard." For example, with a stroke or a traumatic brain injury, if the region of the brain responsible for perceiving the sound is damaged, it can result in deafness, despite fully functioning sensation. Conversely, if the eardrum or small bones are damaged, the brain is still able to perceive sound, but the ear is unable to sense it. In some cases, this can be corrected such as with a cochlear implant. Surgeons can surgically implant the device inside the cochlea, which is connected to an external microphone. The microphone circumvents the eardrum and small bones and transmits the signal directly to the device in the cochlea to restore hearing.

The specialized sensory organ (ear), combined with the specialized perception areas of the brain, allows us to hear. This is essentially the same idea for all of our senses. We have photoreceptors in our eyes to detect light, haptic receptors in our skin to allow us to feel, chemoreceptors in our nose to allow for smell, and taste buds (chemoreceptors) on our tongue to allow us to taste. Specialized corresponding regions of our brain process the information from each type of receptor to formulate our perception of the stimuli. Thus, in order to sense and perceive the thoughts of someone next to our hero, they would need to have both a sensory organ to detect the electrical impulses travelling along their neighbour's neurons and a dedicated area of their brain for interpreting these signals.

Outside of science fiction, there is actually precedent for this in nature. Sharks have a sensory organ on their snouts called the ampullae of Lorenzini that enables them to detect electrical impulses.[28] When fish swim around, their nerves and muscles generate electrical impulses, which sharks can detect and use to track their prey. A Dutch scientist named Adrianus Kalmijn, studying this ability in the 1960s and 1970s, conducted experiments to test shark abilities.[29] Kalmijn and his team first hid fish in tanks with sharks, and the sharks searched out and ate the fish. The researchers then buried electrodes set to simulate the motor activity of the fish, and this stimulated searching and feeding behaviours in the sharks. These electrical impulses are generated from the shark's dinner's movement rather than their thoughts, but this serves as an example of an electrical-sensing organ; if sharks can do it why couldn't a human! Further, there are also additional stimuli all around us that sit outside human perception simply because we are unable to sense them. For example, humans can detect sounds with frequencies between 30–17,000 Hz, whereas dogs can detect sounds between 67–45,000 Hz.[30] If I were to blow on a dog whistle (30,000 Hz), I wouldn't hear anything, but Xavier, currently sleeping next to me, would wake up in a start. Even though there are sounds occurring around us, we are unable to perceive them because we lack the necessary hardware. Similarly, our eyes see light with wavelengths between 400 and 700 nm,[31] but light waves are travelling around us at all kinds of frequencies. When I sit on a beach in the sunshine, I can see the visible light emitted from the Sun, but the red hue my skin develops a few hours later is evidence that there are also ultraviolet waves (<400 nm wavelength), which I can't see, that are hitting me as well. In fact, astronomers have developed telescopes to view space across the light spectrum beyond what our eyes can see. This finally leads us to the question of what other stimuli there are around us that we just don't have either the organ to sense or the ability to perceive, like thoughts.

Super Perception of Thought

In order for our hero to be able to read someone's thoughts, they would need a sensory organ to detect the stimuli and the ability to interpret those signals. Since we have established that thoughts are just the firing of neurons and that these signals can be measured by detecting electrical fields (like sharks do), this ability could come about by being able to detect the electrical activity of neurons.

In the laboratory, we measure firing patterns of neurons through electrodes placed on the skin. If you place an electrode over an individual's femoral nerve (the nerve that innervates (or controls) their quadriceps muscles) and ask them to straighten their knee, you can measure the electrical activity of the nerve as the signal travels from their brain to their muscles. If you place similar electrodes on the skin of their scalp, you can measure neural activity patterns from their brain. If that person were to speak to you, then the language centre of their brain would become electrically active. If you pinch them, emotional and pain centres fire, and if you ask them to wiggle their toes, then the motor regions of their brain become active. Scientists use these techniques to determine which parts of the brain are active and how active they are in response to various stimuli, to better understand how the brain is organized.

In order for our hero to read neuron activity, they would require a sensory organ like a nose, eye, or ear (or an electroreceptor like the ampullae of Lorenzini) that operates like electrodes to sense either ion flow through the neuron membranes or detect the flow along individual neurons. If our hero could detect these patterns of firing (which neurons are firing, how often they fire, and in what order), and transmit this information to the region of their brain dedicated to decoding these stimuli, they would be able to read your thoughts. The simplest way for this to occur would be to have electrode-like organs in our hero's fingers and hands. By placing their hands on their friend's head (like the electrodes I described in the previous paragraph), they would be able to read neuron activity just like we do in the lab, except with much more precision. More useful applications would be the ability to read this activity from a distance by reaching out their hands toward their friend. Even better, this organ could be located within our hero's head (like the shark's ampullae of Lorenzini) and detect all ambient nervous activity, like our other senses do.

As with all senses, it would take time to learn and develop this skill but once developed, it could come as naturally as hearing sounds around us. When you are in a busy coffee shop talking with a friend, you are aware of the noises around you, but you only hear your friend speaking because you are focused on them. Your brain filters out the other noises. If you focus (although it might be rude to ignore your friend), you can then hear all the other sounds and focus on one at a time (depending on how practiced you are at eavesdropping). Similarly, our hero would be able to "hear" or "read" the thoughts of others as background noise. By paying conscious attention, they would be able to focus in and listen to specific sources around them.

Additional Considerations

Telepathic Communication

In many cases, our heroes aren't just able to hear the thoughts of others, they communicate telepathically as well. Like I mentioned earlier, if you place an electrode over the nerve that innervates an individual's quadriceps muscle and ask that person to straighten their leg, you can measure the electrical activity of the nerve. However, if you put a different electrode on their femoral nerve and you run an electrical current across it, you can stimulate that nerve and cause the quadriceps muscles to contract on their own. This can also be done using a powerful magnet over the brain's motor cortex in the specific location controlling knee extension. In the same way, if our hero could electrically stimulate patterns of neurons in the area of their friend's brain that interprets hearing or the nerve that transmits signals from the cochlea, their friend would perceive the sound of our hero speaking to them. Inducing electrical impulses across nerves would require some way to deliver the impulses, which we won't attempt to discuss here but is really just the contrast to reading thoughts.

Conclusion

Our thoughts boil down to patterns of ion flow (electrochemical transduction) along nerve cells. If our hero had both a sensory organ and a perceptual-cognitive process to decode nerve impulses, they would be able to read the thoughts of those around them. Detecting and perceiving these signals would give an individual immense superhuman abilities, which would hopefully be used for more than just eavesdropping!

CHAPTER 4

Laser Vision

Many heroes rely on their abilities to save the day at close range through hand-to-hand combat. However, many others can effectively work from a safe distance using projectiles, utilizing prowess with a bow or various firearms. Distance gives our projectile-wielding hero the advantage of being able to save the day while standing safely back. One particularly effective method is through powerful lasers. Lasers can be emitted from a variety of places on the body, or outside of it, such as a laser rifle or a weapon mounted on a bionic suit (more on that in Chapter 8), but most often our hero emits lasers from their eyes.

Lasers have many practical uses. For example, when I give presentations or lectures, I often use a laser pointer to direct the audience's attention to something on the screen that I would like emphasize or highlight. In the laboratory, we point lasers through tissue samples and record how the lasers are diffracted by the tissue to tell us properties of that sample. Other lasers are used to remove a tattoo or zap away unwanted body hair. Lasers are useful in medical procedures, and some lasers cut or engrave metal and wooden objects. These lasers have commonalities, but there are also key differences. For example, I'm not concerned that my laser pointer will cut through the screen during a lecture or burn a hole through my pocket if I accidentally left it on. Conversely, if you use a laser engraver to remove a tattoo then you're going to remove more than just the tattoo. This chapter will begin with a description of what lasers are and how they work. I'll then speculate on how a laser could be generated within the eye without affecting our hero's normal vision.

Lasers

Fundamentally, lasers are a form of light radiation. Light has some interesting behaviours that scientists are still trying to figure out. Specifically, light has a very unique wave-particle duality, where it behaves like both a wave and a particle. All quantum physics depends on this contradictory duality, even though it is seemingly impossible to think of light as a wave (continuously distributed in space) and a particle (localized in a finite region of space). Light travels at a given wavelength, frequency, and velocity, and it can be refracted, interfered with, and diffracted—all of which are properties of a wave. However, light also exists as having distinct quantities of energy, and carries momentum, both of which are properties of particles. I'm not going to enter a debate over the wave-particle duality of light, so I'll leave it there, but the important takeaways I want to highlight are that light consists of packets of energy quanta called photons and that these photons behave like both particles and waves.

Many of these factors give light its properties. Figure 4.1 depicts a schematic of a simple sinusoidal waveform. In this figure, you can see some key factors about the wave such as wavelength (distance between peaks), frequency (the number of peaks in a given time), and amplitude (height of the wave). Frequency and wavelength are, basically, measures of the same thing, so I'll address them as the same. The wavelength of a light wave dictates the colour we perceive that light to be. For example, light with a wavelength of 450 nanometres (nm) we perceive as blue, 520 nm we see as green,

560 nm as yellow, and 650 nm as red. From most sources, like the Sun, light is emitted broadly across the full light spectrum. Some of this light we can see with our eyes, but there are parts of its spectrum, like infrared and x-rays, that we can't detect. When light hits a solid surface, some of the light is absorbed while the rest is reflected. The colour we perceive objects to be is determined by which wavelengths of light get absorbed, and what wavelengths get reflected. For example, when light comes in contact with an object that we perceive as blue, it absorbs wavelengths above and below 450 nm and reflects 450 nm light. When that reflected 450 nm light reaches our eyes, it stimulates photoreceptors sensitive to that particular wavelength and our brain interprets it as blue. Additionally, the amplitude of the light dictates its energy level or intensity. A low amplitude wave has low energy, and a higher amplitude wave has higher energy. In this way, the wavelength and amplitude of the waveform govern the colour and intensity of the light.

The word laser is an acronym for Light Amplification by Stimulated Emission of Radiation. That is a mouthful and it's clear why we choose to call them lasers, but it gives insight into how a laser works. A simple laser is comprised of three main components: an active medium, a pumping source, and an optical resonator (Figure 4.2). At the heart of a laser is what's called an active medium and is usually a crystal, glass, or gas. This active medium is primarily comprised of an amalgamation of atoms but contains a few larger molecules as well. In case you don't remember the structure of atoms from your biology and chemistry classes, I'll highlight the important characteristics.

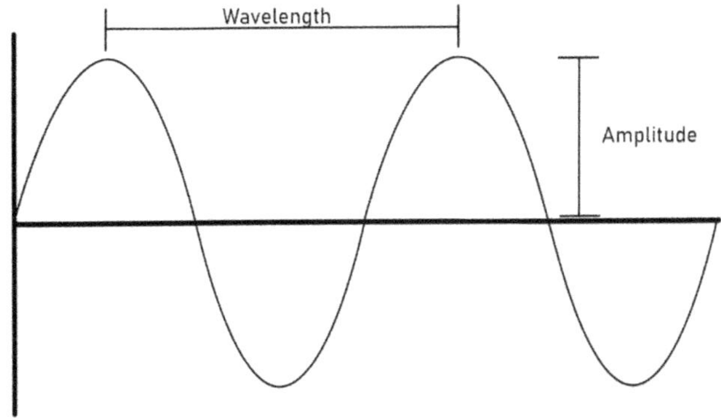

Figure 4.1: Schematic of a simple wave such as
a wave of light and its characteristics.

Atoms are comprised of a nucleus that is orbited by electrons. Laws of electromagnetic attraction organize electrons into specific configurations around the nucleus. This electromagnetic attraction between electrons and the nucleus keeps the electrons more or less at consistent distances from the nucleus. For an electron to move further away from the nucleus, it requires external energy to overcome this attractive force. Energizing the electron causes it to become excited and enables it to temporarily move out of its energy well and further from the nucleus (Figure 4.3). When the electron returns to its original position, it returns (emits) the energy in the form of a photon (light energy). This process of de-excitation is what generates the light (photons) of a laser.

Imagine if you had a heavy bowling ball sitting on the floor next to you. In order to pick that ball up over your head, you would need to exert energy to enable your muscles to produce sufficient force to lift the ball. That ball, once lifted, now has potential (stored) energy and if you let go of the ball (watch your toes!), that potential energy converts into kinetic (movement) energy, and the ball comes crashing back down to the ground with a loud thump. In a similar process, the electron (like your bowling ball) needs to be energized to move away from the nucleus. When de-excitation occurs, the potential energy is released in the form of a photon, rather than kinetic energy and the loud noise the bowling ball makes when it collides with the ground. When de-excitation of an atom is done in synchrony with adjacent atoms, lots of photons can be generated together. The active medium of a laser contains a high proportion of atoms in the higher energy state (with excited electrons), which is achieved by the pumping mechanism.

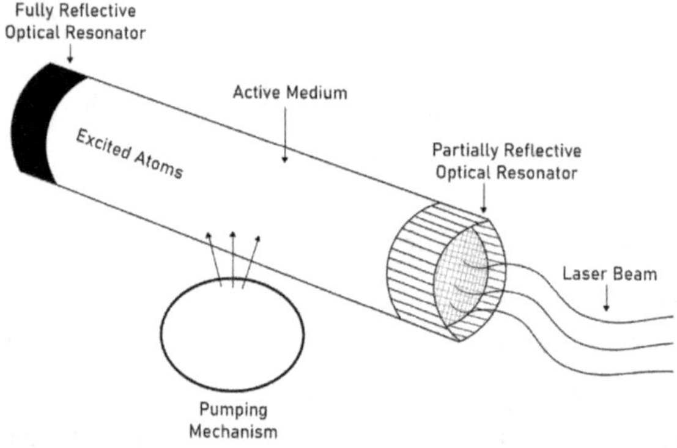

Figure 4.2: Schematic of a simple laser. The pumping mechanism excites atoms within the active medium. Atoms give off photons when de-excited, causing a cascade and synchronized release of photons around the active medium. Photons reflect back and forth between the optical resonators and are emitted from the partially reflective end as a laser beam.

Pumping can be achieved electrically through electrical stimulation to excite electrons, chemically through chemical reactions to excite electrons, or thermally by heating of the atoms to excite electrons, anything that transfers energy to the atom to excite the electron. Once the environment with energized electrons is created, a cascade of photon emission occurs, and electrons become de-excited and release photons. As these photons pass other energized atoms, they stimulate the de-excitation of neighbouring electrons and more photons are released. Each of these photons stimulates the de-excitation of another atom and, in this manner, large quantities of photons are released together. These photons accumulate as they buzz about within the active medium and are directed by the optical resonators. The resonators are on opposite ends of the active medium, and one is 100% reflective to direct the photons back and forth over the medium. The other optical resonator is partially reflective and directs some photons back across the medium but also allows photons to be released as the controlled beam of the laser (Figure 4.2).

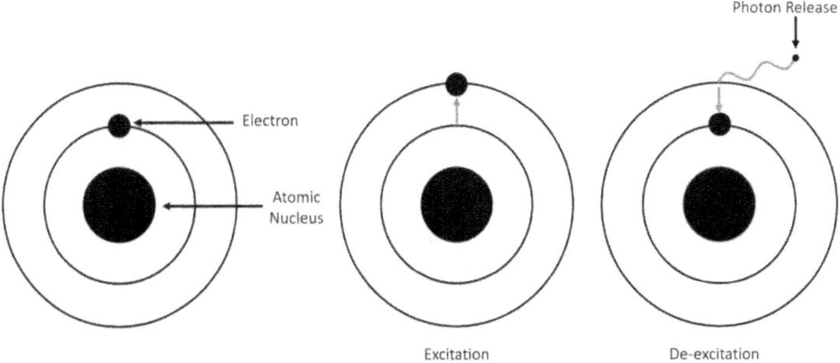

Figure 4.3: Electron excitement and de-excitement. The pumping mechanism provides energy to excite the electron and allows it to move further from the nucleus. When de-excitation is triggered, the electron returns to its original location and emits a photon as energy. It is these photons that compose the laser.

To recap, when a laser is turned on, pumping begins to super energize electrons, electrons become de-excited and release photons, these photons trigger de-excitation from adjacent atoms, and more photons are released in a synchronized cascade. Photons bounce back and forth through the medium creating an intense photon beam, which exits through the partially reflective end of the optical resonator.

Thus, for our hero to possess laser vision, they would need to have a localized region containing the active medium, a way to energize the atoms (pump), and a resonator to control the laser beam release. Next, we will examine the properties of our eyes to speculate where this could occur.

The Eye

The eye contains the physical structures responsible for producing our sense of sight. The eyeball is essentially a fluid-filled sphere with a hole on one end to allow light in, and a series of sensors sensitive to light (photoreceptors) on the opposite end. The main structures at the front of the eye are the pupil, the iris, and the lens, which work in concert to allow light to pass into the eyeball (Figure 4.4). As light approaches the eye, it passes through the superficial layer (cornea) and reaches the pupil (the opening). Depending on how bright it is, the pupil will constrict or dilate to control how much light is able to pass into the eye. This constriction or dilation is achieved by the iris. The iris is what makes your eyes brown, green, or blue, and works as an aperture to control the size of the pupil and how much light can enter the eye. When you go from inside to outside on a bright day, your iris will constrict your pupil to limit the light getting in to keep the inside of your eye from being overexposed to light. If you then enter into a dark room, your iris dilates your pupil to allow more light to enter. Once an appropriate amount of light is allowed into your eye, the lens focuses that light onto the back of your eye. This focusing is called accommodation and ensures you can focus on objects both close and far away. This light, refracted by the lens, travels through the fluid within your eye and terminates on the retina lining the back of the eye.

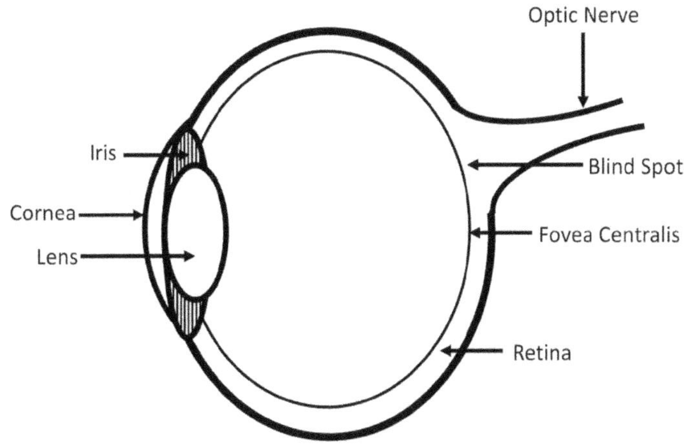

Figure 4.4: Structure of the human eye.

The retinal lining on the back of the eye consists of sensory receptors called rods and cones. Rods and cones are sensitive to light and when stimulated, send signals to the brain for processing (like hearing described in Chapter 3). Cones require a greater amount of stimulation but enable colour vision, while rods detect black and white but have a lower stimulation threshold. Dogs, for example, have a greater proportion of rods compared to cones so Xavier can't see things in the same colour I do. However, he is better able to guard the house against raccoons and shadows at nighttime when we are all trying to sleep.

The location of the stimulated cones or rods on the retina informs where objects are in the field of view, while the frequency (or wavelength) of light stimulating each cone and rod provides information about colour. Directly at the back of the eye is a region called the fovea centralis, the region with the largest density of cones. When you look directly at and focus on something, the fovea centralis is the area primarily responsible for what you are seeing. This high density of cones is why you have the best visual acuity when you look directly at something, and your periphery remains just a little bit blurry. You can only detect light that lands on cones or rods. When the photoreceptors are stimulated, they transmit their signals along neurons on the back of the eye behind the retina and away from the eye via the optic nerve. The optic nerve then carries this signal to the brain for processing. Importantly, where the optic nerve connects to the retina, there are no rods or cones. This means you cannot detect light in that region; it is your blind spot. If you close one eye, there is a small region in your field of view that you are unable to see. With two eyes, these blind spots are in different areas of your field of view, and your brain fills it in with what the other eye sees or ignores them so that you don't notice them. What is important to note about this is that the addition of the optic nerve that connects to the eye takes up space on the back of the retina that would otherwise be dedicated to photoreceptors. There is only so much space on the back of the retina, so with laser vision it is important to consider where the laser is located.

Laser Vision

Given the anatomy of the eye and the way that anatomy functions to allow for sight, there are limited locations that a laser generator could be. Ideally, if an individual wanted to fire a laser from their eyes with maximum accuracy, this laser would be targeted at the location they are focused on. This means it would ideally need to pass from the centre of the retina, travel through the fluid of the eye, through the lens and out the pupil at directly what our hero is looking at. This would allow the hero to have the greatest accuracy as the laser would be targeted at whatever they were focusing their eyes on. However, as with the optic nerve, a laser (active medium, a pumping source, and an optical resonator) would take up space that is otherwise dedicated to rods and cones. Worse yet, this is the location of the fovea centralis. Additionally, with the lens focusing light coming into the eye, a laser passing through the lens would be refracted by the lens and might affect accuracy of where our hero is intending to aim.

A laser contained within the eye would, therefore, need to have three factors. First, it would need the components for a laser: a chamber containing an active medium (most likely a gas), a pumping source to stimulate electron excitement, and the optical resonator to release/focus the laser. Next, it would need to be located outside of the densest regions of the retina. Otherwise, it would significantly impact our hero's vision by taking up space otherwise dedicated to light sensation.

Additionally, since the laser can't be located directly in line with the line of sight, it cannot fire directly in line with where our hero is looking. This means that depending on how far away the target is, the trajectory of the laser needs to be adjusted. When aiming a rifle, for example, a marksman looks through a scope affixed to the top of the gun. When the gun is fired, the bullet travels along the barrel of the gun and is projected in the direction the barrel is pointing. Because the scope cannot be in the exact line of the bullet, as it would block the bullet trajectory, it is located slightly above the barrel and points down slightly to accommodate for the difference between the barrel and sight. As a result, the rifle is only accurate within a range. To compensate for this, the scopes of rifles can be angled up and down to change the range over which the scope corresponds to the bullet's trajectory. In order to fire at the further target, the scope is angled up, and to fire at the closer target, the scope is angled down. In the same way, the trajectory of the laser would need to be adjustable. Thus, for our hero to have laser vision, the design of the laser generator would need to be oblong, located at the back of the eye but not in a particularly dense region of photoreceptors, and would need to be controlled by a small muscle to allow for precise aiming. This would have the smallest effect on the hero's ability to see and would allow them to maintain accuracy on targets set at both close and distant ranges.

Additional Considerations

Laser Power

The ability to generate a laser, like my laser pointer, from within the eye would be quite a feat and would be a handy party trick. However, unless their villain is a house cat, they are unlikely to be foiled by the laser pointer I carry around in my pocket. If an individual wants to use laser vision for feats of heroism, that laser must have sufficient energy to cut, burn, or melt objects. The laser must have an incredibly high frequency, high-energy beam, like those used for laser cutting or engraving.

Energy

Such a high-energy beam would require a substantial amount of energy to generate. The laser's pumping source would need to be incredibly effective since the laws of thermodynamics govern that energy would need to come from somewhere (energy cannot be just created). Thus, our hero would need a mechanism to either store chemically, trigger electrically, or develop heat, to feed their pumping source. Since these resources are limited and/or would need time to be restored, there would be a limit to our hero's laser "fuel" before they would need to "recharge." These lasers would likely then be limited to relatively short bursts and would need time to regenerate.

Conclusion

The back of our laser-emitting hero's eye would need to contain the three major components of a laser: the active medium, the pumping (energizing) source, and the optical resonator. The laser must not be located directly at the back of the eye; otherwise, our hero would have trouble seeing. Accordingly, our hero would need to be able to make small adjustments to the aim of their laser to accurately control the trajectory of the beam. With just a glance, our hero would be able to save the day while maintaining a safe distance from danger.

CHAPTER 5

Shrinking and Growing

"Pick on someone your own size" is a classic superhero quip. Drawing the attention of a villain away from innocent bystanders allows our hero to tackle the villain head-on or keep the villain distracted while their super-buddies pull off some sneaky gambit unnoticed. However, this quip only works if the hero is the same size as the villain. For most of our heroes, this depends on how big (or small) they are and how big the villain is. However, this restriction is not relevant for everyone. The ability to shrink or grow enables our hero to fit (pun very intentional) any situation.

The ability to shrink or grow enables our hero to slip through keyholes, observe as a fly on the wall, essentially disappear from sight, deliver critical blows to their enemies, or make clever quips However, our hero would still be governed by the laws of nature, which, for several reasons, both restrict and enhance these abilities. To begin, let's consider what is actually changing size to allow our hero to shrink or grow.

Natural Growth

At your last physical exam, your doctor likely took a measure of both your height and weight. He or she would have compared them to each other and then to a standardized chart to compare your height and weight to that of your age. The reason doctors make this measure is that humans all have the ability to grow. Simply accelerating these natural processes, however, would be too limited to account for a "super" ability.

I'll use one of the bones of the lower leg (the tibia) as an example. The tibia is the long bone in your shin that articulates at the knee and ankle. The length of the tibia grows throughout infancy, childhood, and adolescence, before slowing and stopping before adulthood. The rates of growth are highest during infancy and puberty. During these phases of growth, the length of the tibia increases (along with the rest of our bones) to allow for us to grow taller. The tibia is categorized as a long bone, meaning it has two ends called the epiphyses and a long segment in the middle called the diaphysis. Before the advent of x-rays, it required a very clever experiment done by a scientist named Stephen Hales to determine that bones grow from their ends.[32,33] Hales drilled holes along the length of long bones in young chickens and measured the distance between the holes as the chickens grew. He identified that the holes near the ends of the long bones grew apart, while holes in the centre of the bone did not. He concluded the bones had to be lengthening from the ends. With more sophisticated technology available to us, we now know that bones grow from what is aptly termed the growth plate, a structure located at the ends of the bone shaft in-between the epiphyses and the diaphysis.[34] To do so, cartilage cells in this region begin to proliferate (we get new cells) and those cells grow, thereby causing the length of the bone to increase. These cartilage cells are then replaced with solid calcified bone. This occurs at both ends of the bone causing the bone to elongate in both directions.

Although this process is quite incredible, it is by no means super. The growth of a tibia from a few centimetres to 30–50 centimetres over 13–17 years is less drastic and far too slow to be of any use against a supervillain. Additionally, this is a one-way street. Once the growth plate closes and is replaced by solid bone, the bone can no longer grow, and it certainly can't shrink back to its original size. Even if it could be reversed, it would have distinct limits on either end and would require concurrent shrinking processes to occur in tandem between the bones, ligaments, blood vessels, nerves, and muscles, all at the same time, at the same rate and magnitude. The natural processes of growth describe nicely how we develop but poorly assist us to vanquish evil. Therefore, the mechanism for our heroes to change size cannot arise from accelerated and/or reversed natural processes of growth.

Human Composition

In order to talk about what is changing in size, we need to understand what we are made of. To start big and work to small: a person is an organism (a living being) and is comprised of organ systems. Organ systems are comprised of individual organs, themselves comprised of tissues, that are comprised of similar types of cells. For example, a person's muscular system is made up of organs called muscles; muscles are formed by muscle, tendon, nerve, and vascular tissues. Muscle tissue in particular is made of muscle cells (fibres). Our cells are made up of organelles, each with unique functions, and these organelles are all made up of molecules. Molecules are the structural arrangement of elements, or atoms. The atom is where I propose the magic happens.

History of Atomic Theory

Discovering the nature and structure of atoms (atomic theory) occurred largely between 1805 and 1930. Much earlier, somewhere around 400 BC, a fellow named Democritus came up with the idea that matter is composed of basic types of particles.[35,36] Essentially, if you cut something in half enough times eventually you cannot cut it in half anymore and what you have left is the fundamental particle, which is indivisible. Democritus was Greek so he called this *Atomos* the Greek word for indivisible. Jumping forward many (over 2000) years, John Dalton noted that if you break down certain molecules, they always result in the same constituents in the same proportions. For example, if you break down a water molecule, it always results in oxygen and hydrogen, and there are always two hydrogen atoms for every one oxygen. In 1805,[37] he reasoned that all substances consist of tiny particles, which he termed atoms (a throwback to Democritus' *Atomos*). Further, he reasoned that all atoms of an element would be identical and that when atoms form molecules, the proportions of elements would be consistent.

If you rub a balloon back and forth on your hair and then lift the balloon away, your hair stands on end and extends toward the balloon. Observation of this phenomenon in the late 1700s led to the hypothesis that electric particles transfer between objects. In fact, as you rub the balloon on your head, electrons (which are negatively charged) transfer from your hair to the balloon and attract the positive charges on your hair, which causes your hair to become attracted to the balloon (negative attracts the positive). It was also observed that the attraction between your hair and the balloon is proportional to the distance between them (how far away the balloon is moved) and the amount of energy transferred (how much you rub the balloon). In 1897, J. J. Thomson was doing experiments using cathode rays, which are essentially neon lights, and determined that electrical charge was able to bend the cathode rays.[38,39] He was able to surmise that the "ray" was actually a stream of negatively charged particles of very small mass, and he thus discovered the electron. He applied his discovery of the electron to John Dalton's idea of atoms and integrated observations of transferring electrical charge (electrons) between objects, to form his model of the atom. As an Englishman, he described his model of the atom as "plum pudding" where the positively charged matrix was like the pudding, and the electrons randomly moving around the matrix were like the raisins (Figure 5.1). When the balloon rubbed against your hair, or the cathode created a ray of energy, it was the raisins (electrons) being transferred between puddings (atomic matrixes).

The discovery of the atomic nucleus is credited to Ernest Rutherford in 1911 who came to the conclusion after interpreting the results of experiments done by his colleagues Hans Geiger and Ernest Marsden.[40] In their experiments, they shot particles through a gold film (gold pressed really, really, really thin) and observed where the particles ended up. What they observed was that almost all the particles were deflected by less than 1 degree (passed almost right through the gold film) (Figure 5.2). However, a very small number were deflected substantially (as much as 90 degrees). Rutherford made some key conclusions from this result.

The fact that the particles were able to pass through the gold film with very little deflection indicated that the gold film was not entirely solid and that particles could pass through the gaps. Additionally, since only a few particles were deflected, the gold film was mostly comprised of gaps, which discredited the plum pudding model. Rutherford concluded that the mass of the atoms, the part that deflected particles, was concentrated within a very, very small area. He termed these packets of mass the atomic nucleus. It was determined quickly after that the atomic nucleus is made up of protons (positively charged) and neutrons (neutrally charged).

Soon after (1913), a Danish physicist named Niels Bohr (who trained under Rutherford) applied principles of planetary motion to further refine the atomic model.[41] Using the hydrogen atom (the hydrogen atom consists of one proton and one electron) as his model, he reasoned that the attraction between the heavy proton and the relatively lighter electron would cause the electron to orbit the proton as the Earth orbits our sun (Figure 5.1). Given the understanding of the laws of motion, this electron would orbit in ellipses with the radius of its orbit dictating its angular velocity and energy level.[42] In the 1920s, this model was furthered with the introduction of quantum mechanics by Werner Heisenberg, Erwin Schrodinger, and Paul Dirac. Even though electrons orbit the nucleus, they travel as both a particle and a wave, where the distance from the nucleus increases and decreases, and instead of travelling in neat elliptical orbits, they move around within electron clouds (orbitals) or regions where they are *likely* to be found (Figure 5.1).

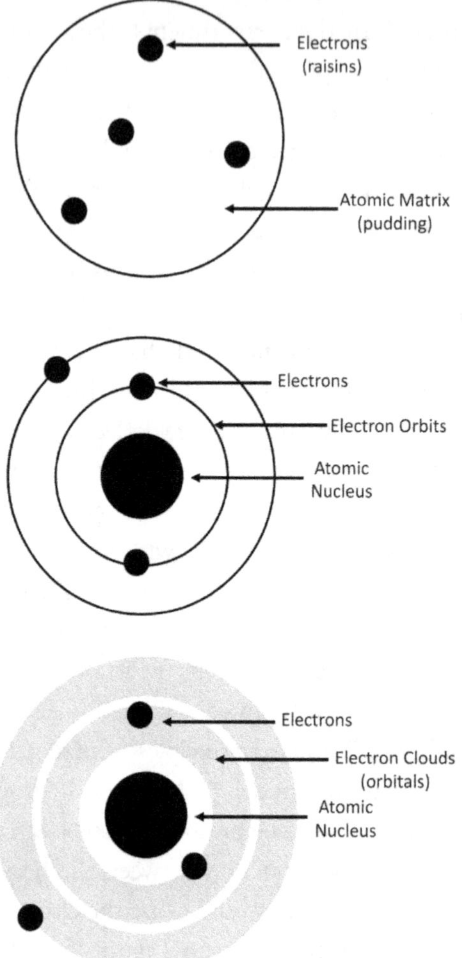

Figure 5.1: Historical developments of the atomic structure. Top image: Thomson's "plum pudding" model where electrons moved through an atomic matrix. Middle image: Rutherford and Bohr's model with atomic nucleus and elliptical electron orbits. Bottom image: Current understanding of atoms, where electrons move through electron clouds.

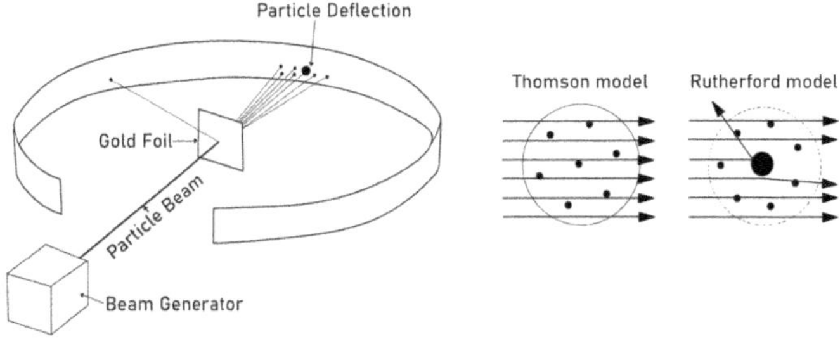

Figure 5.2: Rutherford's experimental setup. A particle beam
was sent through thinly pressed gold film. Most of the particles
went right through the empty spaces between the atoms of
the gold film. Only a few were deflected, indicating the mass
of the atom was condensed in a small space (nucleus).

Atomic Structure

Much of this section has been alluded to in the previous section on
the development of atomic theory, but, for the sake of clarity, I will
summarize and describe in more detail the current understanding
of atomic theory. The atom consists of three elementary particles:
electrons, protons, and neutrons. Electrons are the negatively
charged particles that orbit the atomic nucleus. They are relatively
light compared to protons and neutrons and cannot be broken
down any further. They exist within the electron cloud, attracted
to the nucleus and in constant flux around it. Rather than
travelling in neat elliptical orbits like the planets, we consider the
probability of the electron being in any given location within the
electron cloud and around the nucleus.

The nucleus is the centre of the atom and is comprised
primarily of protons and neutrons. The protons and neutrons are
much (over 1000x) larger than electrons, although neutrons are
slightly heavier than protons. Unlike electrons, both protons and
neutrons are comprised of even smaller particles called quarks.
Protons carry a positive charge, while neutrons have a neutral

charge, and the nucleus as a whole has a positive (opposite to the electron) charge. It is this charge gradient that creates the attraction between electrons and the nucleus. If you look for an element on your periodic table, you will see the atomic mass, which is the sum of the all the masses of the protons, neutrons, electrons, and what is known as the mass defect, but we won't get into that. By definition, the charge on a proton is +1, on an electron is -1, and a neutron is 0. Thus, the number of electrons relative to the number of protons dictates the overall charge of an atom and its ability to interact with other atoms. When this number is not zero, the atom is called an ion and can be an anion (negative) or a cation (positive). Because of this mismatch in charge, anions love to bind with cations. For example, oxygen ions have a charge of -2, meaning the nucleus contains two fewer protons than electrons in orbit. Since hydrogen ions have a charge of +1, two hydrogen ions (1 + 1 = 2) readily bind to a single oxygen (-2) to stably form H_2O, or water.

Given what we know about atomic theory and the structure of our basic particles, we can now surmise what would need to occur for an individual to shrink.

Super-Size Changes

There are three conceivable potential mechanisms for how shrinking could occur: removal of atoms, shrinking of the entire atom, or decreasing the electron orbitals and bringing neighbouring atoms closer together.

Removing atoms would decrease both an individual's mass and volume. There would very simply be less of "them" and they would become smaller. The major issue with this option is if atoms are taken away, they would need to be stored. Matter cannot be destroyed and nor can it suddenly rematerialize or be created (laws of thermodynamics) when our hero wants to regrow. Further, they would need to assemble correctly and in the right order as before.

Additionally, once enough atoms are removed, the function of our cells and tissues would begin to degrade. If you remove too many atoms from a particular position, or even certain key ones, it could be debilitating or even deadly. Thus, removing atoms wouldn't be viable for shrinking. The second option would be to shrink down all of an individual's entire atoms, elementary particles and all. However, since electrons are energy waves/particles we cannot change how big they are. Further, as per their name, atoms are "indivisible" and the splitting of atoms has serious consequences of the nuclear fallout variety.

Since removing or replacing atoms isn't feasible and we can't change the size of our elementary particles, the most reasonable option remaining for our tiny hero is to decrease the space between atoms. The diameter of the nucleus is about 100,000 times smaller than the diameter of the atom. Put another way, if we imagine an atom to be the size of the Earth then the nucleus would be about the size of London's Wembley Stadium. Or if the atom was the size of Wembley stadium, the nucleus would be the size of a tiny aphid. Most of the atom is made up of empty space through which our electrons travel. Simply put, for our hero to shrink, they just need to condense this space. If we halve the volume of the electron cloud without changing the size of the elementary particles, then we would effectively shrink to half the size. Since the elementary particles comprise such a small proportion of the atom's total volume, this gives immense potential. There are, however, consequences to this. The position of electrons around the nucleus are not entirely random. They are governed by charge and attraction between the electrons and nucleus. Additionally, the repulsive forces between the positive charges in neighbouring nuclei would resist the condensing of space. In order to bring the electrons closer to the nucleus, our hero needs to contend with these forces. Therefore, the fundamental mechanism for the condensing of electron clouds has to include the mitigation of these forces. To shrink, our hero would either need to activate a dormant particle somewhere within the atoms to alter these forces or decrease the charges on the protons and increase the charges on electrons (decreasing repulsive charge between neighbouring nuclei and decreasing the electron cloud size). Thus, by decreasing the size of the electron cloud and bringing adjacent atoms closer together, our hero could shrink, or by the opposite, grow. However, it is also worth considering the laws of nature, which both limit and benefit this ability.

Nature's Laws

We live in a universe with laws. We don't fully understand those laws, and they can be interpreted in interesting ways as Albert Einstein did with mass and energy, but they cannot be broken. The laws our tiny heroes are most concerned with are laws of thermodynamics and laws of motion.

Law of Conservation of Mass

One of the laws of thermodynamics, the law of the conservation of mass, was first penned by a fellow named Antoine Lavoisier and states that that mass cannot be created nor destroyed.[43] During chemical reactions, the components of molecules are rearranged to form new compounds, but the total mass of the chemicals entering the reaction is the same as the mass of the new chemicals after the reaction. An object's mass is equivalent to its density multiplied by the volume it takes up in space. Humans have an average density of around 1 g/cm^3: for every cubed centimetre of human, they weigh one gram. This does vary a bit depending on their body composition: fat has a density around 0.9 g/cm^3 and muscle has a density around 1.1 g/cm^3. A muscular individual will be denser than an individual with a greater percentage of body fat. When our hero shrinks, their mass is conserved (it has to be!), but since they take up less space (volume has shrunk), they have become denser.

When our hero shrinks or grows, they are confined to the law of the conservation of mass. This means that if our hero weighs 160 pounds (mass of 72 kg) when they are normal size, then when they shrink down to the size of a bug, their mass is still 72 kg and they still weigh 162 pounds. This also means if our shrunken hero runs across a shelf, that shelf needs to support 72 kg of mass and may be susceptible to breaking. It is also worth considering pressure. If you lay down on a bed of nails, you can actually be quite comfortable. This is because your body mass is divided by the surface area of all the nails, and the pressure exerted on your back by each nail is relatively small. If, however, you lay on one nail, then all your body mass is exerted on the very small surface area of the single nail and this high pressure becomes very uncomfortable. In the same way, if our hero shrinks down, the surface area they contact the ground with is very small and the pressure they exert on the ground becomes quite high. This runs the risk of breaking what they stand on. On the positive side, if our tiny hero decides to hit someone or something, they can do so with substantial force as indicated by our laws of motion.

Laws of Motion

Isaac Newton formulated three laws based on his observations of how things move through space. These laws of motion conclude that momentum, like mass, must be conserved after a collision. Momentum is the product of an object's mass and the velocity at which it travels. The conservation of momentum dictates that if two objects collide, we can calculate how fast they will travel and in which direction after the collision. This also has profound implications for our tiny hero. If our full-sized hero takes a running leap and while in the air they shrink, their momentum is conserved, because their mass is conserved. If our hero takes a running shrinking leap that happens to be directed at their nemesis, then they will collide with the villain with the same momentum as they took off with. This allows our hero to pack a serious punch. Instead of getting hit by a fist-sized fist, imagine getting hit with the force of a 160-pound marble. Now you can understand how our villain feels!

Additional Considerations

Magnitude of Change

The main consideration for our protagonist is the magnitude of change in their size. There are two primary limiting factors to how much our hero can shrink: the size of elementary particles and the relationship between body volume and surface area. First off, decreasing the attraction between adjacent nuclei and shrinking the size of the electron cloud can only go so far. Because the elementary particles can't be shrunk, our hero could never become smaller than the combined volumes of all the nuclei. Additionally, even if the orbitals were shrunk to almost nothing, the density (the same mass within a very, very small volume) would become overwhelming. In fact, a large mass occupying an infinitely small volume results in a black-hole-swallowing-the-universe-type situation.

The mathematical relationship between volume and area is simple but has significant implications for shrinking. To keep it simple, let's imagine a cube as our example. If the length of the cube is doubled, then the area of each side is quadrupled, and the overall volume of the cube is increased by eight times. Because area scales with a power of two (squared) and volume scales with a power of three (cubed), small changes in length can cause profound changes in area and volume. In particular, this affects the sensitive ratios between surface area and volume. Many of our organs depend on these volume-surface area relationships. For example, the stomach, intestines, blood vessels, and lungs, among others, rely heavily on these ratios for their normal function (more on this in Chapter 9). Without additional compensation, sustained or middle to long-term shrinking can only occur within the functional limits these systems rely on for volume and surface area.

Interacting with the World Around

During inhalation, air is drawn into the lungs. On Earth, that air is around 21% oxygen. The oxygen molecules enter your lungs, diffuse into the blood, and attach to hemoglobin molecules circulating your bloodstream. The thickness of the walls of the lungs, the size of hemoglobin molecules (on red blood cells), and the diameters of your blood vessels are the size they are because of the size of oxygen molecules. If our hero shrinks, then these transport systems (lungs, hemoglobin, blood vessels) will all shrink as well. If red blood cells are too small for oxygen to bind to them, then they will be unable to carry oxygen throughout the body. More drastically, if our hero shrinks down so much that oxygen molecules can't enter the lungs or are too large to diffuse through the lungs into the blood, our hero will suffocate. It is unreasonable to expect our hero to hold their breath for the duration they are shrunken, so either the hero needs a supply of shrunken oxygen like a little scuba diver, or the shrinking effect must be transferable to the air immediately around them.

This morning after my workout, I had a protein smoothie with yoghurt that contained prebiotics. Prebiotics are insoluble types of fibre; they serve as food for the trillions of micro bacteria that live in the gut and assist with digestion. If I were able to exert a shrinking effect on the atoms of my body but failed to exert that effect on the bacteria living in my gut (and elsewhere), I would encounter some very unusual sensations as the trillions of microorganisms suddenly grow from my insides! Thus, our hero's shrinking ability would have to be exerted on the atoms of their body, oxygen that they breathe, and the microorganisms, food, and foreign bodies that may be within them at the time.

Conclusion

The basic unit of structure, the atom, is largely "empty" space. The condensing of this space by manipulating the elementary particle charge could allow these atoms to shrink. The ability to shrink gives our hero the ability to dodge, hide, or deliver effective blows. However, our hero's abilities are limited by natural laws and the ability to maintain tissue and organ function. Additionally, this effect must be applied to foreign invaders to their body and the oxygen they need to breathe.

CHAPTER 6

Teleportation

Imagine you are a supervillain, and you are trying to rob a bank with explosives and a giant drill. You are about to get started when you turn around and your explosives are disarmed. You turn around again, and your drill bit is missing. Then you turn around again, and your explosives are gone altogether! You haven't seen anything and as you ponder what is happening, you get struck from behind. You turn around while falling, to see what hit you, and you get struck again from the other side. The next thing you know, you've been teleported outside the bank and into the back of a police car. Your plans have been foiled by a teleporting hero! Teleportation allows our hero to get in close, deliver a blow or rescue their love-interest (or both!), and then teleport back to relative safety. The ability to teleport from one physical location to another has incredible utility.

There are two distinct ways our hero would be able to instantaneously travel through space. First, our hero might be able to bend space, allowing them to travel from location to location. Second, our hero might be able to dissolve their body and reappear in a new location. Since these are two distinct mechanisms, I will address them separately.

Bending Space

Travelling through outer space is a very exciting avenue for science fiction. To paraphrase the intrepid Captains Kirk and Picard, it opens up the possibility of exploring strange new worlds and seeking out new life and new civilizations. However, the biggest limiting factor for celestial travel is the time it takes to travel from location to location. As an example, our sun is 150 million km from Earth.[44] At this distance, it takes the Sun's heat and light energy, which travel at light speed (300 000 km/second), 8 minutes and 20 seconds to reach Earth. In relation to the infinite expanse of space, Earth and it's sun are right next to each other. Since we aren't able to travel at the speed of light, travelling outside our solar system and back for a holiday using conventional rocket propulsion is impossible. Kirk and Picard's USS Enterprises get around this issue by using Zefram Cochrane's warp engine. The theoretical premise of the warp engine is that the warp field bends space and allows the starship to take shortcuts across the expanse of outer space. In reality, we often think about space as an empty three-dimensional vacuum where our position can be determined with coordinates (x, y, z) and we can move linearly around it, but it seems that really isn't correct. Based largely on the theory of relativity, we understand space to exist in gravitational waves similar to those of the ocean. If we imagine space as a flat plane as in the left image of Figure 6.1, we like to imagine travelling from point A to B using path 1. However, our understanding of quantum physics dictates that space is actually bent as waves (gravitational waves) like in the right of Figure 6.1. In order to travel between points A and B, we need to follow path 2 (the same distance as path 1). What the warp drive does is allows the spacecraft to move between the waves and take path 3, drastically shortening travel time. But let's move away from outer space travel and return to our superhero.

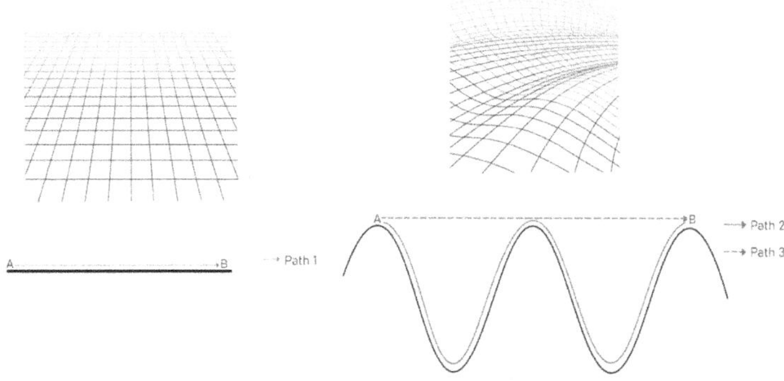

Figure 6.1: Depiction of travelling through space under two different paradigms. The left image depicts traditional understanding of travelling linearly from point A to B via path 1. The right image captures current understanding of gravitational waves. Travelling through space is like sailing a ship up and down the waves (path 2). However, the ability to travel across the wave peaks (path 3) would enable the traveller to "teleport" through space.

Leaving behind science fiction and space exploration, and returning to our superheroes, the same idea of bending space (this time not outer space) can be applied. Let's imagine our hero standing and facing their villain as in Figure 6.2. Normally, in order to get behind the villain, our hero would need to run from point A up to and around the villain to point B (path 1). However, if our hero were able to bend space, it would be a simple matter of taking path 2 from point A to point B, which to the villain would seem like our hero has disappeared and then reappeared behind them. Our hero would have effectively teleported from point A to point B where they can deal with the villain.

There are two possibilities to utilizing this method of teleportation. First, space is bent until it becomes close enough for our hero to pass between the two ends. Second, our hero would be able to bridge the gaps between the existing folds in space. Or, a combination of the two where space is bent, and a bridge is formed to close the remaining gap allowing our hero to pass across.

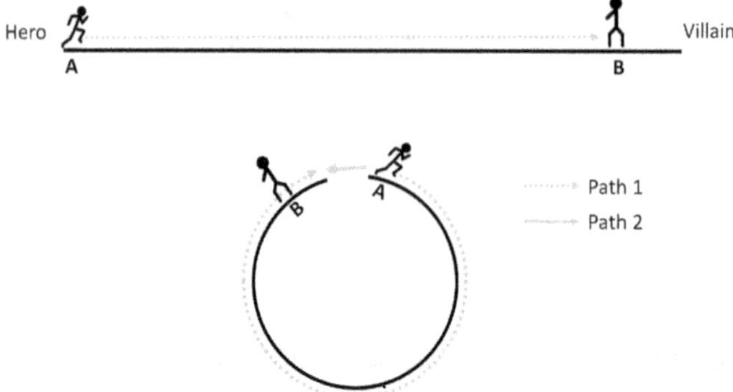

Figure 6.2: Teleportation through bended space. Top image depicts our hero travelling normally toward their nemesis (path 1). Bottom image depicts how our hero, having bent (or curved) space, can travel traditionally along path 1 or by teleporting across bent space via path 2.

Now, it is of utmost importance that our hero has control over the ability to open, hold open, and close their portals. The worst-case scenario would be for space to unbend while our hero is partly through the portal. Half of our hero being in one place and the other half in another when the portal closes spells catastrophe. Our hero also needs to be able to unbend space to close their portal, in order to prevent unwanted tailgaters or stowaways from following them.

Matter-Energy

Albert Einstein must have suspected his work would have a major impact on the field of theoretical physics, but I don't think he would have anticipated that his theories would appear in a book on theoretical physiology. Nevertheless, the paradigm shift he instituted on natural laws is the focus here. In the previous chapter,

I described Antoine Lavoisier's law of the conservation of mass: mass cannot be created nor destroyed.[43] Our understanding of this law was deepened by Einstein who described the energy component in this relationship.[45] He reasoned that matter is comprised of both energy and mass with his famous equation $E = mc^2$. E is energy, m is mass, and c is the speed of light. Mathematically, because energy and mass are on opposite sides of the equation, they can be converted back and forth by involving speed (light speed, specifically). Therefore, things (mass) can be converted to energy, and energy into things. As a bit of an aside, the idea of the big bang is that infinitely dense energy at the centre of the universe converted into the mass (stuff) that has been strewn throughout the universe.[46] For now, even though mass must be conserved, it can exist as either mass or energy.

Let's put it together for our hero. Teleportation in this instance consists of the breaking down of their tissues, converting them into energy, transporting that energy to a new location, converting the energy back into mass, and then reassembling our hero as a whole. A few key points to this method are that our hero isn't exactly teleporting but transporting as energy. This has three major implications. First, energy isn't really a "thing," so it has to travel via a medium. For example, the Sun's energy travels to us on Earth as heat and light energy. The implication here is that if energy is transmitted by a medium such as light, energy is unable to travel faster than the speed of light. Thus, our hero can only travel between locations as fast as light speed. Second, when an energy "signal" is transmitted, there is potential for that transmission to be interfered with, which would have devastating results for our hero. You may be concerned that they would be unable to teleport through walls since visible light cannot travel through objects, as evidenced by shadows. However, some light, x-rays or gamma rays, for example, can travel through objects, like your body at the doctor's office or your suitcase at airport security. Remember also, from Chapter 3 and 4, that not all wavelengths of light can be detected by our rods and cones. If our hero were to travel at certain wavelengths (outside the visible light spectrum) they would be invisible to the human eye. If our hero were to transmit their energy as light at the right wavelengths, they could travel at light speed, would not be visible during transport, and could travel through walls.

Finally, disassembling and reassembling our hero is an infinitely complex process. The previous chapter dedicated a great deal of attention to the atomic composition of our physiology. I won't rehash that now, except to reiterate the levels of structure from the bottom to the top. Starting at the bottom we have the atoms. Atoms are comprised of elementary particles, where the number and proportion of elementary particles in each atom govern which element the atom makes up. Elements combine with each other to form molecules. Molecules form together into compounds that are assembled into amino acids, which form chains to create proteins. The order of amino acids in the chain is dictated by our DNA and governs the function of the protein. These proteins form together to create the organelles and structures of the cells, as well as the intercellular (between the cells) environments. Cells of similar types form together to create tissues. Like tissues form organs that operate in concert within organ systems. These organs perform our body's functions and make up the whole organism. Our hero has to be put back exactly (exactly!) the way they were prior to the teleport, and ideally, this should occur as quickly as possible.

Additional Considerations

Reaching the Destination

Teleporting your body somewhere else involves the reappearance of your physical body elsewhere. One of the biggest risks to our hero is reappearing in physical space that someone or something already occupies. For example, if our hero converts their mass to energy, and then forms back to mass in the middle of a wall, there will be conflict between their mass and the wall in that space, a conflict that isn't likely to benefit them or the wall! Additionally, if they teleport blindly or they miss their target, they may find themselves surrounded by their villain's henchmen, falling, or in the presence of some other danger (lack of oxygen, hot or cold temperature, projectiles, etc.) that they didn't anticipate.

When marksmen fire a gun at a target, they need to consider the effect that gravity has on the bullet and adjust their aim accordingly. When the target is placed at a sufficient distance, they also need to contest with the Coriolis effect. Even though the bullet travels in a straight line (assuming wind is minimal and accounting for gravity as just mentioned), the target moves with the rotation of the Earth and to the orbit around the Sun. I think it is safe to assume that when our hero converts their mass to energy, that energy will travel in a straight(ish) path. But, if they don't aim their energy stream, with the orbit or rotation of the Earth in mind, especially when covering long distances, they may end up floating around in outer space!

Conserving Momentum

If I take a running start and jump through a doorway, I emerge in the other room with the same speed and moving in the same direction as I had entered the door, thanks to the conservation of my momentum. When our hero travels through the bridge or portal connecting bent space, it's logical to assume the same thing happens. If our hero jumps backwards through a bridge, they will emerge on the other side travelling backwards. However, this would also mean they would be unable to teleport without moving as they would need to move to pass through the portal between the two locations. The exception being that unless the portal were to open beneath their feet and they drop through, they would be technically falling through the portal. If our hero uses the second method of teleporting, when their mass is reformed from energy, it's logical to assume they would coalesce stationary as there is no momentum to be conserved.

Conclusion

Transmitting an individual through space has incredible utility. From dodging or delivering blows in a fight, getting to and saving citizens or fellow heroes, or cutting down on the commute to the superhero's lair, the ability is very appealing. This can be accomplished either through bending space to create portals between locations or converting their mass to energy and back again. Either way, they had better know what's on the other end or they may end up somewhere worse off than where they left!

CHAPTER 7

Super Healing

Saving the world from supervillains is a dangerous business. Our heroes regularly get punched, kicked, bit, electrocuted, thrown, shot at, burned, frozen, and the list goes on and on. For most of our heroes, they need time to recover. They need to "RICE" (rest, ice, compress, and elevate) their wounds, massage their stiff muscles, let their cuts and bruises heal, and take some time off for rest and relaxation. They may even have an orthopaedic surgeon (like my wife) on speed-dial. Fortunately for these heroes, they would most likely have Band-Aids with the logos or faces of their superfriends on them. However, there are a select few heroes who don't have the same use for icepacks, Band-Aids, or physiotherapists because their abilities include super healing.

Healing

The body is under a constant flux of breaking down its tissues (catabolism) and rebuilding them (anabolism). The combination of these processes (metabolism) can determine the health of many of our tissues. For example, muscle undergoes continuous muscle protein breakdown and muscle protein synthesis. If muscle protein synthesis exceeds that of breakdown, then the muscle grows, and conversely, if the breakdown exceeds synthesis, then the muscle atrophies (shrinks). For example, factors like resistance exercise training and protein intake boost muscle protein synthesis, which is why our muscles grow with training. Because these processes are so dynamic, this can occur very rapidly. In another example, a study following seven healthy young men confined to bedrest for just 6 weeks, reported 25–30% loss of lower limb strength and 15% loss in thigh muscle cross-sectional area.[47] In a similar fashion, bone undergoes continuous breakdown and regeneration. Due to many factors with aging, when this breakdown exceeds repair, osteoporosis or osteopenia (weak bones) occur. Therefore, balancing the metabolism of our tissues is an important factor for maintaining health while our tissues undergo repair. This is particularly true following an injury. This chapter will look at three examples of tissue healing: cuts to our skin, healing of a broken bone, and regrowth of a limb. The first two occur in humans already (sometimes with help from stitches, casts, or bone plates) and the latter occurs in select few animals.

Skin Healing

Skin consists of two primary layers: the outer epidermis and the dermal connective tissue that underlies it. The epidermis is the protective outer layer that acts as a barrier between you and the outside world. Beneath the epidermis lies the dermal layer, which consists of connective tissues, hair follicles, and blood supply. Several sequential and overlapping processes are involved with the healing of skin, and they begin immediately after a cut is sustained. Because your skin acts as a barrier to the outside world, damage to the skin must be quickly addressed to prevent microbes or foreign bodies from entering and causing infection. This healing process occurs through four main phases: clotting, inflammation, regeneration, and maturation.[48]

Clotting

When skin is damaged deeper than the epidermal layer, the small blood vessels running through the dermal layer are cut and release blood into the wound. In response, the ends of the blood vessels constrict to minimize further blood loss. Blood contains several components including red blood cells (carry oxygen), white blood cells (manage infection), platelets (clotting agents), and other chemicals (hormones, growth factors, and proteins). During normal circulation platelets, which are responsible for the clotting of the wound, are inactive, so they flow freely through blood vessels. At the wound site, blood exposed to air triggers a cascade of reactions resulting in the activation of platelets. Activated platelets adhere to each other and the surfaces around the wound. The amalgamation of platelets blocks the severed vessels to prevent bleeding and forms the fibrous structure that becomes the scaffold for the rest of the repair phase.

Inflammation

Once bleeding is under control, the inflammation phase begins. The amalgamation of platelets into a fibrous structure not only serves as a temporary barrier closing off the wound, but the platelet amalgamation also embeds chemical messengers called cytokines and white blood cells within itself. These white blood cells are the immediate response of the immune system to protect the body by eliminating foreign bacteria and break down damaged cells into spare parts. Pro-inflammatory cytokines are released to attract more and more white blood cells to the site of the injury. The chemical messaging that attracts white blood cells is the cause of swelling, as blood flow to the area increases to facilitate delivery of white blood cells to combat infection and to remove cellular debris. At the end of this phase, the fibrous matrix has been laid down primarily by platelets; white blood cells have cleared away damaged cells and foreign bacteria; and pro-inflammatory cytokines have attracted the factors required for constructing new tissues.

Regeneration

In undamaged skin, the epidermis replaces itself every 30–40 days by breaking down old tissue and rebuilding the epidermal layer.[49] The cells responsible for forming new skin cells are called fibroblasts and reside primarily in the dermal layer. During the inflammation phase, fibroblasts are also attracted to the wound site by growth factors and cytokines that are released. Fibroblasts arrive on the scene and begin to synthesize connective tissue (collagen) to replace the temporary fibrous scaffold initially laid down by the aggregation of platelets. Along with the collagen scaffold, blood vessels are synthesized to allow for the building blocks and energy needed for the synthesis of new tissue to be delivered through the blood.

Maturation

The final phase of the healing process involves the synthesis of new skin (scar) and the closing over of the wound. A stronger form of collagen is laid down over the scaffold and scar tissue replaces where the skin used to be before the injury. The wound begins to heal starting at the ends of the cut and working toward the middle until the ends meet and close over fully with a scar.

Process

These four phases are not temporally distinct, and they overlap each other to form the whole healing process. The time-course for full healing of a wound is dependent upon many factors such as the health status of the individual and the depth or length of the wound. Most skin wounds heal after 21 days, but only regain around 80% of the original skin's tensile strength.[50] Additionally, external factors can affect the rate of healing. For example, uncontrolled diabetes affects the action of white blood cells, which slows the inflammation process; stress affects cytokine responsiveness; smoking decreases blood flow to the skin; and genetic disorders like hemophilia affect platelet action.[51–53] If healing can be slowed by these factors, then it is conceivable that it could be sped up by other factors (super factors, perhaps), but we will come back to that.

Bone Healing

The healing process of a broken bone is similar to skin, but there are some unique differences.[54,55] Namely, unlike skin healing where the original skin is replaced with scar tissue, bone is healed with bone and the end product is a full regeneration of the tissue's structure and function. The processes of clotting and inflammation occur as they do in skin following injury. The clot forms from platelets, the platelets form a fibrous scaffold and release pro-inflammatory cytokines (chemical messengers) to attract white blood cells, the site is cleaned, and damaged tissue is broken down. The differences in healing commence in the regeneration phase and involve mesenchymal stem cells. Mesenchymal stem cells are specialized cells with the ability to differentiate (transform) into many types of mature cells such as bone, muscle, cartilage, and fat. In the regeneration phase of bone healing, rather than collagen being laid down over the scaffolding, mesenchymal stem cells arrive and are triggered to form osteoblasts (cells that form bone) and chondroblasts (cells that form cartilage). Initially, chondroblasts begin to lay down cartilage over the site, filling in the scaffolding. Capillaries fill the gaps through the soft cartilage until the blood supply is diffuse. The diffuse blood supply allows for the propagation of osteoblasts that begin to weave in bone. In the maturation phase, bone eventually takes over the cartilage and begins to calcify and harden until it has achieved the structural and functional characteristics identical to the bone before the injury.

The most distinct difference in the healing processes between skin and bone is the involvement of mesenchymal stem cells. Stem cells, in general, are cells that, when triggered by the right stimulus, can develop into different types of cells. For example, hematopoietic stem cells are the precursors to the different kinds of blood cells. When they are triggered, and depending on what signal they receive, they differentiate into the cell required. Mesenchymal stem cells, as mentioned previously, can differentiate into bone, fat, muscle, or cartilage. Since stem cells are not involved in the skin healing process, skin is replaced by collagenous scar tissue. However, bone healing utilizes stem cells to close and repair the wound with replacement bone tissue. To take this healing further, let's consider the regrowth of a limb.

Limb Regrowth

For obvious reasons, the ability to regrow limbs has fascinated scientists for over 250 years.[56] However, the advancement of laboratory techniques has made its study substantially more effective in recent years. Much of this work has been conducted on newts or salamanders, and in particular, a species of salamander called the axolotl has been studied predominantly.[57,58] If a vigilante crime-fighting axolotl gets their forelimb cut off by a villainous amphibian, the limb regrows and, incredibly, the new limb is a perfect duplicate, indistinguishable from the former limb.[58] If the left forearm is cut off but then implanted into the right limb, the resultant limb, once healed, is a right limb with both a right and left hand. Further, if a forelimb is amputated at the wrist (hand cut off) and then the wrist stump is grafted to the animal's side so both ends of the arm are attached to the animal, and then the elbow is severed, new arms and hands grow out of both ends that were formerly the elbow.[59] Finally, if the nerve of the right limb is damaged and a piece of the left limb is transplanted near the damaged nerve of the right limb, the grafted piece grows into an extra limb sticking out of the right arm. Study of this remarkable process of healing has revealed three phases: the initial injury and immune action, re-epithelialization, and regeneration.

Injury Immune Action

The first step in regrowing a limb is to have it amputated. This sounds very obvious, but there are important events that arise from the injury itself. As in skin and bone injuries, clotting and inflammation comprise the hemostatic response. The injured site receives the same white blood cell response to ensure foreign pathogens don't gain access to the inside of the body and to break down and repurpose damaged cells. Also, the epidermal cells at the edge of the injury begin to replicate and migrate over the cut surface. Within 12 hours after the amputation, this epidermis has grown to cover the surface of the stub with a thin layer of skin-like epithelium.[60]

Re-epithelialization

The cells within the thin layer of epidermis then undergo cell division to multiply and form the thicker multi-layered epithelial cap. This cap protects the accumulation of mesenchymal cells, which have been recovered from around the site and are travelling from adjacent areas. This accumulation of mesenchymal stem cells forms the blastema, a cone-shaped cluster of undifferentiated stem cells at the base of the limb stub that becomes the site and source of regeneration.

Regeneration

Cells within the blastema differentiate into the target cells (muscle, bone, nerve), which are used as the material to rebuild an exact duplicate of the severed limb from the base of the severed limb up to the tips of the brand-new fingers.

Super Healing

If our super-healing hero is on the receiving end of a beat down and sustains a broken arm from a well-placed attack, they need to have that arm healed fully and back to full strength before the next flurry of attacks. They don't have the luxury of waiting 6–8 weeks for the bone to reach its full strength. Similarly, if our hero receives a particularly nasty injury and they heal within seconds of sustaining it, it won't do to have lasting deficits of function. Thus, whenever super healing is portrayed, it generally has two dominant features: a full restoration back to original structure and function, and it occurs at a greatly accelerated rate.

Healing Quality

Skin healing can be a useful model for describing the mechanisms required for full healing. As I described previously, adult skin heals with the use of scar tissue and only heals to about 80% of its initial tensile strength.[50] What I didn't mention is that if an embryo sustains a cut, the embryonic skin heals perfectly without developing a scar.[61] The fact that skin heals perfectly on an embryo, but the same skin once born heals with a scar, provides a useful model of the same structure undergoing two different healing processes. Scientists have studied this phenomenon using laboratory animals by surgically cutting embryonic skin and observing the adaptive responses.[62] They have identified that embryonic skin exhibits a lower inflammatory response and has exposure to greater amounts of growth factors that are already present as a result of skin growth occurring in development. The result of these differences is that in non-embryonic skin, collagen is laid down in disorganized patterns as scar tissue and does not regain its previous integrity. Conversely, embryonic skin is replaced with original materials to a complete restoration of the structure and function of the skin.

In different models of healing, bone and axolotl limbs regenerate to their original capacity through the use of stem cells. The advantage of stem cells is that they can become cells of the target tissues needing repair. The limitation of stem cells is they require a complex signal to initiate differentiation. Stem cells have been explored in healing capacities in humans to limited success mainly for this reason. For example, osteoarthritis is a condition characterized by degeneration of cartilage in joints, leading to painful bone-on-bone contact. Scientists have attempted to treat arthritis by injecting stem cells into the joint capsule with the intent that the stem cells will differentiate into cartilage and restore the cartilage surfaces.[63] This is highly intuitive, but unfortunately, the mere presence of stem cells in the joint is insufficient to stimulate new growth and this is thought to be because the injected stem cells don't receive the correct signals. For stem cells to be effective, they need to be present at the correct location, in sufficient amounts, and need to receive the right signal to differentiate into what is required. So, for our hero to heal back to full capacity, they would require substantial anabolic growth factors to stimulate tissue regrowth, sufficient quantities of stem cells available to replace old tissues, and the signals required to stimulate stem cell differentiation into any of the required tissues.

Healing Rate

Skin and bones usually take around 3–4 weeks and 6–8 weeks, respectively, to heal. The incredible process of limb regeneration can be accomplished in as little as 7 weeks.[57] The fact that our bodies repair themselves like they do is remarkable all on its own. But when we discuss super healing, these processes are largely inadequate.

Assuming the previous section is correct and stem cells are involved in our hero's healing, there are a lot of processes that need to occur in a short time-frame. As a sample, once a wound has been sustained: clotting, an inflammatory response to remove damaged tissue, an accumulation of stem cells and stimulatory growth factors, and the signals for what to develop, all need to occur at the injury site. Additionally, the stem cells need to undergo differentiation and replication to produce the replacement cells required, which then need to be assembled into tissues. All these factors, though many of them can occur simultaneously, need to be incredibly accelerated.

At the opposite end of the spectrum, healing disorders can be affected by a single step. Hemophilia patients, for example, lack a single clotting factor in their blood and, so, when they sustain a cut, the clotting phase is severely affected. This creates a bottleneck and resultant backup in the healing process and leads to about 20% longer healing compared to non-hemophiliacs.[53] However, if a single process or factor is accelerated, rather than impaired, it has a much smaller effect. For example, if clotting were to be instantaneous following a cut, it would accelerate the overall healing time, but the wound would still need to progress through the inflammation, regeneration, and maturation phases. Thus, for healing to occur rapidly, every step needs to be accelerated. There really isn't a single mechanism that could lead to this accelerated wound healing. Every step in the chain would need to be accelerated or it will be limited by the weakest link or links. Overall, our hero needs to have an incredibly high metabolism to sustain all these accelerated processes and would need to effectively store stem cells and growth factors throughout their body to be readily available when required.

Additional Considerations

Injury Severity

Regardless of how quickly the healing processes can be accelerated, more severe injuries would take longer to heal. A deep cut would take longer than a shallow cut, a broken bone usually takes longer to heal than a cut muscle, and a severed limb would take even longer. They may all be incredibly accelerated, but these would all take time. Super healing also does not mean invincibility. If our hero sustains an injury too severe to a vital organ or system, or too much blood is lost faster than it can be regenerated, our hero may tragically succumb to their injuries.

Protein Requirements

The average person has a daily requirement of 0.8 g of protein per kilogram of body mass.[64] This protein is required for the regular reparative processes of tissue breakdown and synthesis. Proteins are the structural components that comprise our tissues. Complex proteins are made up of chains of amino acids that form all the organic structures of our body. When we eat protein, our body breaks it down into its constituent amino acids. These amino acids are then delivered throughout the body where they are reformed into chains that form the new proteins (building blocks) for our tissues. When we do resistance exercise training, our damaged muscles send signals that they need to be regenerated and this is accomplished by the delivery of amino acids to the muscle where they are formed into new muscle proteins. Thus, individuals who do resistance exercise training have higher protein requirements, and recent evidence suggests 1.6 g of protein per kilogram of body mass per day is required for those participating in exercise training.[64] This doubling of protein requirements is necessary to repair the muscle tissue damaged during training. Relatively speaking, exercise-induced muscle damage, even following a particularly challenging workout, is nothing compared to the loss of a limb, or a broken bone. To rapidly repair a broken bone or quickly regrow a limb, our hero would need incredible amounts of amino acids to provide the building blocks for the new tissues and replacing stem cell stores. Our heroes had better like meat, dairy, and/or a wide range of legumes, and they should probably invest in a high-quality shaker bottle for their whey protein shakes. Fortunately, like Band-Aids, they could probably get a shaker bottle with their logo or one of their superfriends' logos.

Energy

Each step described above requires energy: production of white blood cells, growth factors, and stem cells, differentiation of stem cells, cell division, and synthesis of new tissues. When this process is distilled into the time-frames that are required for our heroes, the energy requirement is immense. This energy needs to come from food. Along with increased protein requirements, our hero would also need to take in a substantial diet on the days when super healing would need to take place. Fortunately, the enhanced protein intake that would be required is best accompanied by delicious carbohydrates and fats.

Pain

Simply, injuries hurt. Sensory nerve endings run through most of the tissues in our body. When those tissues are damaged, so are the sensory nerves that tell the brain about the damage. This is exceptionally useful to inform us of harmful things, so we can avoid them in the future. From an evolutionary standpoint, this has allowed us to avoid harm, and for me to survive long enough to write this book and you to read it. Day to day, we experience physical pain fairly frequently, and that's even without facing off against villains. Now, even though our hero heals quickly, they would still experience the pain of the injury. It may not last as long because healing of the nerve endings would set in, but pain is pain. Our hero may be able to heal and recover from even grievous wounds, but they would still be best off dodging serious attacks because they are still going to hurt!

Conclusion

Super healing, as we would imagine it, involves the rapid and full restoration of the structure and function of our tissues. This requires an enhanced ability to synthesize and circulate growth factors, stem cells, and amino acids, and to do so at largely accelerated rates. However, our hero is going to need to consume a rich diet high in calories and particularly focused on protein intake, and their injuries are still going to hurt. As with normal healing, preventative medicine goes a long way: it will be best for our heroes to dodge when they can.

CHAPTER 8

Bionics

In the previous chapter, I highlighted the incredible healing capacity of the axolotl that can regenerate fully functional limbs indistinguishable from the original limb. In some cases, our heroes have lost limbs because vigilante crime fighting is a dangerous profession. Retirement never really seems to work for our heroes, and since very few have the ability for super healing, these heroes either continue facing their villains without a limb or find a replacement limb. In many cases, bionic limbs endow them with additional or enhanced abilities. In a few cases, some of our favourite heroes are just ordinary people who benefit from either bionic or cybernetic enhancements. These range from a single limb to large parts of their bodies to full exoskeletal suits. I think this chapter will be particularly interesting because this is the ability that most readily will make the transition from science fiction to science. Instead of how *it could* be possible, this chapter deals more with what is currently possible. For the sake of clarity, the two main categories of bionic enhancements are prosthetics and exoskeletons.

Prosthetics

Prosthetics have been used since at least as far back as 3000 BC. Egyptian mummies dated to around 2500–2700 BC have been uncovered with carved wooden toe and leg prostheses.[65] There are earlier references in the Indian poem *Rig Veda*,[66] believed to have been written as early as 3500 BC, which refers to a warrior queen injured in battle and fitted with a prosthetic leg to allow her to return to the conflict. There are several other examples from antiquity but for 2000–5000 years, these prosthetics remained largely unchanged. They were generally formed of wood, metal, and leather and were incredibly cumbersome, uncomfortable, and heavy. Minor advancements were made during the Renaissance and the scientific revolution of the Enlightenment, but design of prosthetics really began to substantially advance as demand increased during and following the major world conflicts of the twentieth century. Major advancements since have included movable parts, impact absorption, and modern materials (aluminum, carbon fibre, and electronics).

Original work in prosthetic biomechanics was done to restore partial function to individuals without limbs or parts of limbs. Researchers sought, for example, to allow individuals who lost a leg to walk/limp again or individuals born without fully formed hands to be able to grasp and hold simple objects. Research interests expanded greatly, as is often the case in science, through the advancement of new technology. Juxtaposing the limb used by Terry Fox in his 1980 Marathon of Hope and the limbs used by Paralympians today is evidence enough of these advancements. The rapid development of these technological progressions has led to some interesting dilemmas.

A landmark case came from a South African athlete named Oscar Pistorius. Oscar was born without key bones of the lower legs and had bilateral amputations below both knees. As an adult, he gained the nickname "Blade Runner" because of the pair of J-shaped spring prosthetics he used to run and compete. Oscar was a talented athlete and became highly decorated in Paralympic sport. Given his race times, he argued he should be permitted to qualify to compete against able-bodied athletes at the International

Association of Athletics Federation (IAAF) world championships and the Olympics.[67] The controversy came as a debate over whether replacing metabolically active lower limbs with elastic springs provided him with an advantage over able-bodied athletes. For the first time, prosthetics had advanced far enough that they were considered not only equal to, but possibly even better than a natural limb. I will leave it to the sport sociologists and the IAAF whether this is true, but the fact that prosthetics may actually enhance abilities greatly intrigues many scientists.

The utility of a prosthetic requires both function and comfort. The prosthetic must allow for the individual to perform the tasks of an intact limb, and the human-prosthetic interface must not cause discomfort. The focus of this discussion will be on function, and we will assume our hero is relatively comfortable with their bionic limb(s).

To summarize previous chapters: when an individual wants to move a limb, the brain generates a signal, that signal transmits along the nerves to the muscle, the signal reaches the muscle and triggers it to activate, produce force and shorten. The function of a bionic prosthetic, therefore, relies on the ability of the prosthetic to produce force and move, and the user's ability to control it.

Joint Force

Oscar Pistorius utilized springs in his blade prosthetics to store and return elastic energy. While he was running, when a blade came in contact with the ground, the blade would bend and elastic energy would be stored in the spring. As he transferred his weight over the limb and was ready to take off and propel into the next step, the spring would unbend to convert the stored potential energy back into kinetic energy to propel him forward. In an intact limb, as the foot is planted during running, much of this energy is absorbed as shock through the limb, but the calf muscles contract, and some of this energy is partly stored in the Achilles tendon. During liftoff, the Achilles tendon rebounds to restore this energy and contributes to the propulsive force produced by the calf muscles. In Chapter 2, I mentioned that within a given range kangaroos exert the same amount of energy to cover a wide range of speeds, and this high economy is a result of the energy storage and return of their Achilles tendon. Because so much energy is returned, their leg muscles need to do very little work as they hop along. This is the primary rationale behind the IAAF's decision not to allow Oscar Pistorius to compete against able-bodied athletes, as the elastic storage and return of his blades took over the work done by the leg muscles in those without springs on their legs. For the spring blades to do so, energy must be stored, held, and then returned at the right time to propel the individual in the correct direction. If too much energy is lost, the system becomes inefficient. Additionally, if energy is returned too soon, the runner will be propelled up instead of forward, and if it is too late then the runner won't be able to harness the energy for successive steps.

Using a spring to store and return energy is a very passive mechanism. Other research has been done using active mechanisms for propulsion. Prosthetics with powered actuators have been developed that not only store and release energy but produce force during the propulsion phase of the step. While the individual prepares to lift the prosthetic foot, a motor situated about the ankle activates and generates torque at the ankle to contribute to step propulsion. Step by step, this motor simulates the action of the calf muscles to contribute to walking. This has two major advantages: the motor does the work instead of the individual and the potential force output of the motor could vastly exceed the force output of the calf muscles. Research has shown that prosthetics with motor actuators largely increase locomotor economy compared to spring-only prosthetics.[68] The addition of a motor creates the opportunity to generate force, and potentially enhance joint strength and movement velocity, but we will come back to that. Before the user can take advantage of motor actuators, the user must be able to control the prosthetic.

Prosthetic Control

Early prosthetics were, basically, sticks strapped to people's legs and acted as a rigid extension of the limb. As technology advanced, these were more cleverly shaped and produced of lighter, more pliable materials. However, the user still lacked the ability to control the configuration or position of the prosthesis. The prosthetic may have been modelled to look like a hand, but the fingers couldn't move, and the individual couldn't grab and release objects. This was overcome initially with a clever design utilizing a harness and pulley system to allow grip control (Figure 8.1). For an amputation of the forearm, the prosthetic would be attached to the stub and a cable would be run up the arm and anchored about the shoulder.[69] When the individual flexed the shoulder, the cable would pull and the "hand" (griper is probably more accurate) would be caused to close. When the shoulder was extended again, the grip would release. These systems allowed for control over grasping but were limited because they were weak, they required complex movements of the shoulder and upper arm, and control was awkward, jerky, and difficult.

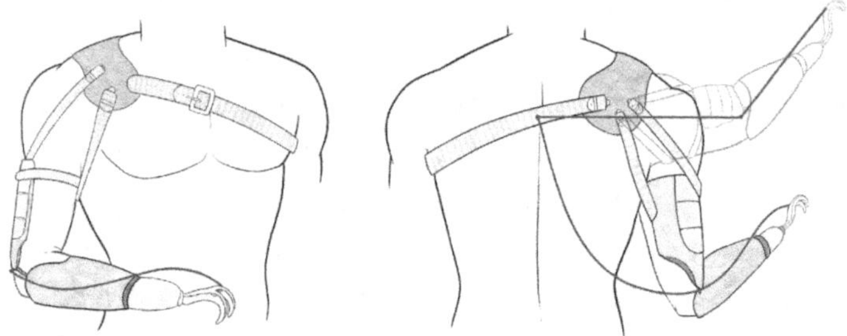

Figure 8.1: Pulley-based prosthetic device. Flexing the shoulder pulls on the pulleys and causes the "hand" to grip shut. Extending the shoulder back down releases the grip.

Major advancements came through the inclusion of what is termed myoelectric integration. These prosthetics interpret what the individual intends to do and transfers those intentions into movements. In Chapter 3, I outlined the transmission of signals via nerves. When the brain intends to move a limb, a signal is generated from the motor cortex, transmitted down the spinal cord, out the spinal root, along the motor neuron, and when it reaches the muscle, it communicates for the muscle to activate. If an arm undergoes an amputation at the elbow, the nervous system upstream remains intact. Thus, if the individual intends to activate their missing hand muscles, the signal is still generated in the motor cortex, transmitted down the spinal cord, and passes along the nerve until it is interrupted by the amputation.

Myoelectric (myo = muscle, electric = nerve signals) integration receives these nerve signals and transmits them to a small computer device. For example, the major muscles involved in flexing the elbow are the brachialis and the biceps brachii; both are innervated by the musculocutaneous nerve. By placing recording electrodes over the musculocutaneous nerve and connecting the signal to a motor generating force at a prosthetic elbow, signals from the nerve can be used to control the motor. This allows the individual to use the natural nervous impulses generated by their brain to move the prosthetic arm as their own. Similarly, by placing an electrode over the radial nerve, which innervates the triceps, the individual could also gain control of extending the elbow. This was first accomplished in 1957 by the USSR Central Institute of Prosthesis and Artificial Limb Making.[70] They developed a motor-powered forearm prosthetic that was made of plastic and ended in a hand-shaped clamp (Figure 8.2). Sensors were placed at the end of the stump and when electrical impulses intended for the "phantom" forearm muscles were detected, the motor caused the hand to close. In this model, the fingers acted all together against the thumb to allow for grasping. This was initially utilized for on-off type control where the motors were either on or off and moved at a constant speed and force. Refinement through proportional myoelectric control has allowed for not only detecting the presence of signals but also interpreting the strength of those signals. The system determines the level of activation of the nerve and translates this level of activation into control over the speed of movement and level of force production. Clever scientists have refined this idea to isolate and distinguish between signals for the elbow flexors/extensors, wrist flexors/extensors/supinator/pronators, and even finger flexion/extension,[71] allowing for much more precise control over the limb. This requires more sophisticated electrodes and, in many cases, the electrodes are fine wires that are implanted under the skin to detect the correct source more precisely. Most of these prostheses record the electrical signals close to or at the interface between the prosthetic and the stub. However, since these signals are generated by the brain and travel down specific tracts of the spinal cord, these signals can be recorded anywhere along the chain.

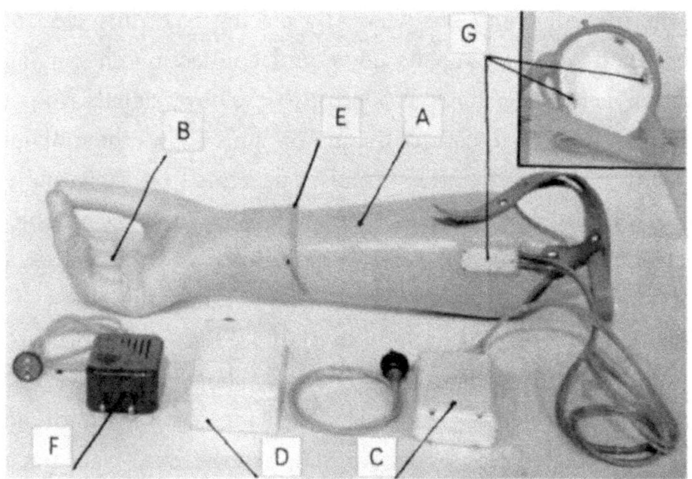

Figure 8.2: Bioelectric controlled prosthetic developed by Popov.[70] Electrical activity of the nerves was detected and used to activate the hand of the prosthetic. Current models have been advanced considerably. Figure reprinted with permission.

An exciting avenue at the frontier of this area of research is the application of this technology to tetraplegic patients. For example, recent work was done by researchers in Ohio using a 53-year-old tetraplegic paralyzed man who had suffered a cervical (neck) spinal cord injury.[72] Spinal cord injuries result in a disruption of the nervous impulses at the level of the insult. When the individual attempts to move their arm, the motor cortex creates the signals, and they are transmitted down the brainstem. Due to the injury, they are interrupted at the level of the injured spinal cord and never reach the muscles of the arm. Thus, because of the injury, the patient lost nervous control of his arm. However, downstream of the spinal cord injury, the nerves and muscles are still functional. For the study, he was fitted with electrodes placed on his head over the motor cortex of his brain and these electrodes were connected to a robotic arm. Incredibly, the subject was able to use the robotic arm to perform both single and multi-joint arm movements to an accuracy of 80–100% in target acquisitions. The fine control over the robotic arm using natural controls is an exciting proof of concept. In essence, this process would be used to regain the function of any, or in extreme cases all, limbs and would result in functioning bionic limbs controlled by the individual's motor cortex.

The nervous system contains two primary functional divisions: the efferent and afferent pathways. The efferent system sends signals from the brain out to the periphery. This largely consists of motor functions to control muscles and are the signals detected by the bioelectric sensors. Conversely, the afferent pathways work in the opposite direction and deliver sensory information back to the brain. If, for example, you grab an egg, tactile feedback from your fingers helps you achieve a grip that is sufficient to keep you from dropping the egg but not so strong that you break the egg. Without sensory feedback, individuals are prone to either dropping objects or applying more force than is necessary. Biomechanists are working on integrating sensory feedback into their prosthetic designs and some devices now utilize microprocessors to determine forces and joint angles.[71]

Think about when you are going down a set of stairs and there is one less step than you thought there was. The floor suddenly interrupts your intended movement and you stumble. Because your foot senses the floor and sends a signal to your brain that your movement has been interrupted, you have a chance to correct, and you may just stumble. Without that sensory input, you would fall head over heels. This is also particularly useful for maintaining posture. A lower-leg prosthesis that can detect its position and auto-correct postural sway can mimic subconscious postural control and allow the individual to maintain balance without direct focus on their limbs. Research integrating sensory-like control to the motors is in early phases and currently has low practicality, but it does show promise.[73] Without a doubt, in order for our superhero to perform at the required level, their bionic limb would need to have sensory feedback about position, and ideally, even provide tactile (touch) stimulus.

Super Implications of Prosthetics

Current bionic prostheses are controlled by the body's natural motor control pathways, allow for fine motor control of the fingers, and can be used to both handle delicate items gently or produce high levels of grasping force. These are not perfect, are expensive, and require further refinement. However, with the rapid advancement of technology and dedication of researchers, I think we can expect prosthetics that function at the same level as an intact unaffected human limb in the near future.

However impressive these prostheses already are, they do not reach the level of performance we would expect a bionic-enabled superhero to have. As design refinements and technology continue to advance, limbs will be able to have many enhanced features. Motors that control the limbs easily have the potential to vastly outpower human muscles and move at much greater velocities. Precision programs could control incredibly precise and stable movements better than the steadiest of human hands. It is unlikely that we will find adamantium or vibranium, like our fictional heroes, but using metal or synthetic materials to produce lightweight limbs able to withstand incredibly high forces and provide protection to the wearer is conceivable. Thus, prosthetics have the potential to vastly exceed the abilities in speed, strength, mobility, precision, protection, and efficiency of a human limb. Bionic limbs also have an additional potential advantage of mechanical enhancements including weapons such as projectiles or lasers, or gadgets such as grapples or thrusters. Replacing one body part, or many, with bionic components has seemingly endless potential.

Exoskeletons

Bionic prosthetics have come a long way and continue to open exciting new avenues of application. An extension of this would be the bionic exoskeleton, which thus far has been largely developed for medical, military, and occupational domains.[74]

A practical demonstration of this is the production of boots similar to the powered lower limb prosthetics described previously. These motors are applied to a boot fitted over the ankle and contribute to the propulsive force during walking or running. Research led by Hugh Herr, a prominent researcher in this field, has led to development of powered ankle orthotics that increase ankle power and improve movement economy, allowing the wearer to walk around while consuming less energy than without the exoskeleton boot.[75,76] Most recently, the development of unpowered exoskeleton ankle orthotics has been underway with mixed success but promising potential.[77] Much of this work has been done with clinical applications in mind such as allowing individuals with gait abnormalities to walk more naturally and controlled. For example, the ReWalk exoskeleton has been designed to allow users paralyzed by spinal cord injuries to walk.[78] The exoskeleton provides external control of the hips and knees and enables the user to stand up, walk at speeds up to 2.2 km/hour, and sit back down again. Further systems have been developed to restore the ability to walk to patients affected by cerebral palsy, stroke, spinal cord injuries, traumatic brain injuries, or even frailty.[79,80] Many of these systems also incorporate the bioelectric control systems used for prostheses to enable user control.

Major advancements to exoskeletons have been achieved through funding dedicated under the umbrella of the United States' Defense Advanced Research Projects Agency (DARPA).[79] This funding has enabled some incredible work. At the forefront of this work was the Berkeley Lower Extremity Exoskeleton (BLEEX) developed under the direction of Homayoon Kazerooni.[81] The goal of this work was to develop a portable robotic exoskeleton to enable soldiers to carry heavy loads. The BLEEX system contains sensors to detect the wearer's movements and controlled motor actuators to move the exoskeleton joints. This allows the load to be transmitted through the robotic exoskeleton rather than the wearer, enabling the wearer to carry a 75-pound backpack almost effortlessly over long distances. Similarly, a company named Sarcos utilized DARPA funding to develop an exoskeleton suit that supported both the arms and legs to not only carry loads but lift and manipulate them as well.[82] Extensions of these and other projects have led to the integration of these systems into the workplace. For example, the FORTIS system developed by Lockheed Martin enables shipyard workers and those in heavy manufacturing to almost effortlessly operate extremely heavy hand-held power tools, thereby increasing efficiency, effectiveness, and preventing injury to workers.[83] There are also many other private companies developing models, but specific information is proprietary and much less readily available. Recently, even exoskeleton gloves have been developed that endow the user with externally controlled fine motor skills.[84] As with all technology, we are likely to see better and better applications introduced into the market. I suspect even by the time you are reading this, even better models have been developed.

Super Implications of Exoskeletons

Current exoskeleton designs enhance the abilities of their operators but are limited by their weight, joint ranges of motion, lag time between sensing and responding to the operator, portability, and battery life. For example, an exoskeleton may provide an individual with the strength of a forklift but if it needs to be plugged into a wall to work, then our hero wouldn't be very effective. However, exoskeletons can already enable people to lift incredible loads and carry them long distances with minimal exertion. The applications of this could easily result in super strength, super speed, greater efficiency, and perfect robotic movement precision. In comparison to the prosthesis, the exoskeleton is particularly effective as it provides protection to the individual inside and with certain materials could provide practical invulnerability to the wearer like a modern suit of armour. Exoskeletons also have the advantage of being able to easily incorporate mechanical enhancements like weapons, wings, or propulsion. Major obstacles still need to be overcome including power sources, mobility, responsiveness to the user, weight, and range of motion. It is most likely that we will see exoskeletons more similar to those faced by Sigourney Weaver's characters in *Alien*[85] and *Avatar*[86] than those operated by Robert Downey Jr. in *Iron Man*.[87] However, the production of an enhanced exoskeleton suit would enable its wearer to safely perform superhuman tasks effortlessly and precisely.

Conclusion

In antiquity, those without limbs were supported by peg-legs that allowed them to stand upright and limp along with the help of a cane. Millenia later, a real-life hero named Terry Fox ran a marathon a day for 143 straight days on a rigid prosthetic with limited shock absorption (not to mention cancer in his lungs!). Decades later, a rigorous set of experiments were conducted to determine whether the blade prosthetics of Oscar Pistorius actually gave him an unfair advantage over able-bodied athletes. Today, current models of bionic prostheses and exoskeletons enable humans to perform limited superhuman tasks. With the further incorporation of stronger and lighter materials and integration of refined control systems, the potential of bionic enhancements to exceed the abilities of human limbs enters the realm of inevitability.

CHAPTER 9

Underwater Breathing

Most of our heroes face their villains on land or in the sky. However, around 70% of Earth consists of water. Oceans are broad and deep and if a villain really wanted to get up to no good, a perfect place to hide their misgivings is under the water where our heroes would be unable to follow. That is, of course, unless our hero can also survive underwater. The ability to breathe underwater allows our heroes to continue the fight where many others cannot follow, and gives them many practical avenues for hiding, travelling, and vanquishing their foes.

This chapter will consider energy and metabolism, which in animals, is tied to oxygen. Then I'll use two examples of respiratory physiology: mammals breathing air and fish "breathing" water to explore how our hero may be able to accomplish both.

Oxygen and Energy

In all living organisms, sustaining life requires energy. This energy is required for locomotion (flight, walking, slithering, swimming), development, growth, healing, maintaining tissue homeostasis, and regulating cellular function. Even if you were to lay in a bed all day watching your favourite superhero movies or reading your favourite comics (or this book over and over) you would still require between 800–2000 calories of energy to keep your body alive, depending on your size.[88] In plants, and maybe select few animals,[89] this energy comes from the Sun through the process of photosynthesis. In animals, this energy is consumed as food and transformed into useable cellular energy in the form of ATP. There are three pathways animals rely on to convert food energy into ATP: the phosphocreatine system, anaerobic glycolysis, and oxidative phosphorylation.

It takes energy to form chemical bonds and join atoms together into molecules (remember Chapter 5) and when the bonds are broken and molecules become atoms, the bonding energy is liberated. Energy for our bodies is provided by breaking apart the chemical bonds of our food, storing that binding energy as bonds between phosphate molecules in ATP, and then breaking those bonds when our tissues need energy. Our bodies require energy for all cellular functions: transferring ions across membranes, moving our muscles, allowing the heart to beat, repairing bones, or fighting infections. Your body is able to glean energy from the three macronutrients we eat: proteins, carbohydrates, and fats. The processes of obtaining energy from proteins and fats are to more or less break them down and convert them into sugars, so for brevity I'll focus on carbohydrates.

Carbohydrates are formed of long chains of simple sugars called glucose. A glucose molecule is comprised of 6 carbon atoms, 12 hydrogen atoms, and 6 oxygen atoms. If you add those up, there are quite a lot of bonds in a glucose molecule. This is why it serves as an excellent source of energy. It is also why sugar has so many calories! These bonds are systematically broken down through the process of digestion. At the very beginning of the process, long chains of sugars are broken down in your mouth, both mechanically as you grind them between your teeth and chemically as enzymes in your saliva break apart the chains. Once you swallow, the chemical process continues through your stomach and into your intestines. Glucose molecules then pass through your intestines into your bloodstream and are delivered all throughout your body where a hormone called insulin helps store glucose within your cells. Breaking down of glucose molecules occurs within the cells themselves.

The first step in breaking down glucose is a process called glycolysis (glycol = sugar; lysis = breakdown). The whole process of glycolysis consists of a series of complicated biochemical steps that break the glucose molecule down into two parts. During this process, hydrogen atoms, consisting of one proton and one electron, are released. The electron is removed from the hydrogen atom and used to bind phosphate to ADP to form ATP, leaving just the hydrogen ion (H+) (really just a proton) to be recycled. Each of the two halves of the original glucose molecule is known as pyruvate and can be used for two purposes. If there is oxygen available within the cell, then the pyruvate is converted to a molecule called acetyl CoA and undergoes a longer series of breakdowns (Figure 9.1).

This series of breakdowns is called the Krebs cycle and exposes the molecule to a series of enzymes that break apart the molecule atom by atom, liberating electrons from the bonds. These electrons pass into the mitochondria where, through interactions with oxygen, they are used to bind loose phosphate ions to ADP molecules and form large amounts of ATP. This energy pathway, the process of binding electrons to oxygen to form ATP molecules, is called oxidative phosphorylation, or aerobic metabolism. This is an efficient method of producing ATP in that it produces the greatest number of ATP per glucose molecule, but it requires oxygen, and it is slow. In some instances, such as during heavy exercise or sprints (when it's hard to catch your breath), you may either need faster ATP production than aerobic metabolism can supply, or oxygen gets all used up and the process gets backlogged. Thus, when pyruvate is formed from glucose and insufficient oxygen is present, the Krebs cycle and oxidative phosphorylation (aerobic metabolism) cannot proceed. Since the cell still needs energy, it must be produced in another manner. Anaerobic metabolism (in the absence of oxygen) is accomplished by partially breaking down pyruvate into lactate and releasing some of the energy from the molecule. This system is much less efficient in that it yields less overall ATP from each glucose molecule, but it occurs very rapidly and can occur in the absence of oxygen. The downside to this method is it also produces a substantial accumulation of hydrogen ions and creates an acidic environment that affects cellular functions. Your body is able to buffer this acidity but not before you feel it as the burning sensation in your legs when you run up a flight or two of stairs. With either method, your body liberates the energy held in the bonds of glucose molecules and uses the energy to produce ATP molecules that are stored within the cells of the body for use when energy is required.

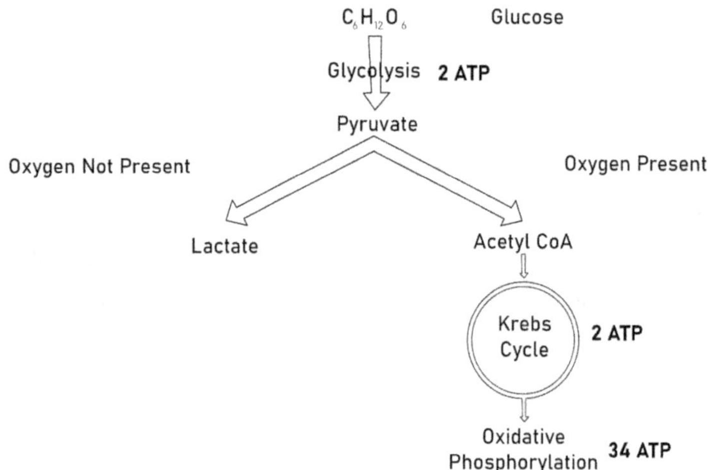

Figure 9.1: Aerobic and anaerobic metabolic pathways. Following glycolysis, if oxygen is present, pyruvate continues through the Krebs cycle and oxidative phosphorylation. Without oxygen present, pyruvate is converted to lactate.

The third energy system is the phosphocreatine system. The phosphocreatine system is a short-term storage system and powers very quick and explosive movements but is quickly exhausted. When you are sitting in a chair you are consuming energy at a low rate. However, if you were to stand up very quickly and start running, your energy demands would suddenly exceed your supply. To compensate, phosphocreatine stored within your muscles splits into creatine and a phosphate molecule. That phosphate molecule forms with ADP, and your ATP supply is partially restored. However, this is only effective in the very short term (approximately 10 seconds) because each phosphocreatine molecule can be used only once and there are limited stores.

Since the anaerobic pathway is inefficient and yields small amounts of ATP and the phosphocreatine system has very limited ATP capacity, it is impractical for sustaining complex life. There are some organisms, bacteria for example, which survive exclusively on anaerobic metabolism and don't require oxygen to survive.[90] However, for the vast majority of animals, the ability to take in oxygen is a requirement for producing the amounts of energy needed to sustain life.

Human Respiration

Humans, like all mammals, provide the oxygen necessary for aerobic metabolism by breathing the air around them. Air on Earth consists of ~79% nitrogen, 21% oxygen, and 0.3% carbon dioxide. These proportions undergo minor fluctuations, if, for example, you stand downwind of car exhaust or live in a polluted city, but they are largely consistent. Air pressure affects the amount of oxygen in the air and thus how easy it is to breathe (why it is more difficult to breathe at a high altitude than at sea level), but the proportion is always constant. The process of breathing takes air into the lungs, removes oxygen out of that air, and exhales the air back out.

Breathing Anatomy

The respiratory system consists of the tubes that transport air within your body. Inhaling draws air in through your mouth and nose and down through the first tube, the trachea (Figure 9.2). The trachea lies immediately behind your esophagus and carries air back and forth up and down your neck. The trachea continues until about halfway down the chest where it splits into two bronchi, the left bronchus that leads to the left lung and the right, to the right lung. These bronchi split into hundreds of smaller branches called bronchioles, which terminate at the alveoli. Alveoli are small balloon-like extensions of the bronchioles with thin (one cell thick) walls. Lining these walls are tiny blood vessels called capillaries. Capillary walls are also ultra-thin (one cell thick), and the inside of the vessels are only thick enough to allow the passage of single red blood cells at a time. Once oxygen reaches the alveoli, it diffuses through the single-cell walls from the inside of the alveoli into the blood of the capillaries.

The basis of extracting oxygen relies heavily on optimizing surface area. By splitting into smaller and smaller branches and ending in the tiny balloon-like alveoli, the inside of the lungs exposes the maximum amount of air possible to the membranous walls, and in close proximity to the capillaries which allows for oxygen to diffuse into the blood. The main constituents of blood (remember Chapter 7) are the red blood cells, white blood cells, and plasma. Very little oxygen can dissolve in the plasma, so oxygen is transported through the blood by attaching to red blood cells.

Red blood cells contain an iron-based protein called hemoglobin, which can bind up to four oxygen molecules per blood cell. Red blood cells transport oxygen through the blood where it releases oxygen to diffuse into the tissues that need it. A number of years ago, high-profile cyclists were implicated in a wide-spread doping scandal involving a drug called erythropoietin (EPO). EPO functions to increase the number of red blood cells in an individual's blood. In turn, this allows more oxygen to be carried and enhances oxygen delivery during exercise. It can also be very dangerous because increasing the number of red blood cells in blood makes the blood more viscous and can lead to heart failure, which is why it is a banned substance. Interestingly, racehorses can store red blood cells in their spleen and release them when they begin to exercise.[91] This natural method of blood doping allows horses to reach higher levels of running performance by increasing the oxygen-carrying capacity of their blood under the demand of exercise (a natural form of blood doping). While oxygen is diffusing from the alveoli into the blood, carbon dioxide is diffusing from the blood into the alveoli. Air within the lungs is exhaled and new air is inhaled, meaning carbon dioxide is removed from the body and new oxygen is brought in. Ambient air that is inhaled consists of around 21% oxygen and 0.3% carbon dioxide and air that is exhaled contains nearly 16% oxygen and 4–6% carbon dioxide.

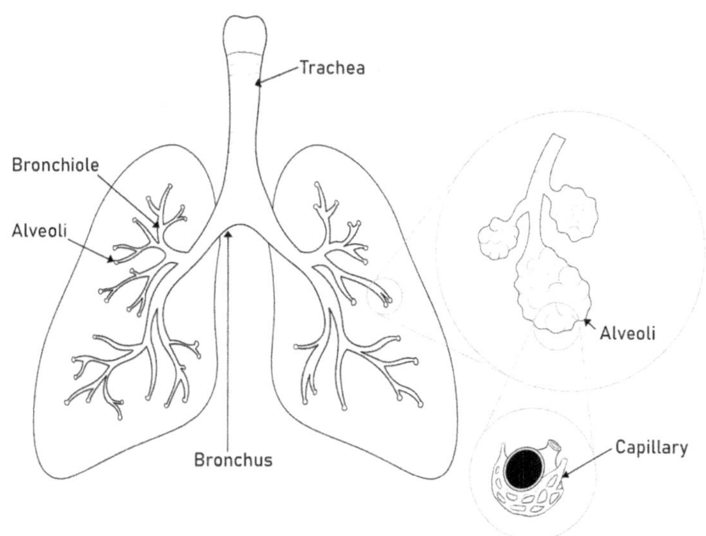

Figure 9.2: Structure and anatomy of the human respiratory system.

The act of breathing is governed by the brainstem. This is automatically controlled and even though you can override and assume conscious control of your inhaling and exhaling (which you have now done because you are thinking about it), your brain does a good job at maintaining your breath at all times. The brain does so by sending a signal down your spinal cord, along a nerve to the inspiratory and expiratory muscles. Breathing is achieved through pressure gradients. Think about a balloon. When a balloon is full, the pressure within the balloon exceeds the pressure outside the balloon; there is a gradient (differential) of pressure. When you release the end of the balloon, the air is forced from the high-pressure space inside the balloon to the low pressure of the room, and the balloon is propelled off like a clumsy rocket. Your lungs are a bit like balloons, hopefully except for the rocket part. Your brain controls the diaphragm, a membrane-like muscle, which spans the bottom of the ribcage and separates the chest cavity from the abdomen. When the diaphragm contracts the whole of the membrane pulls down and creates a negative pressure within the chest (lungs included). This negative pressure acts like a vacuum and since there is lower pressure within the lungs than there is in the ambient air, air comes rushing in and fills the lungs. As the lungs fill, the chest cavity expands, and gas exchange occurs within the alveoli. The expansion of the chest cavity is important because there are small ligaments and muscles that lie in-between your ribs, and this expansion causes them to stretch like elastics. When your diaphragm relaxes and abdominal pressure drops, the stretched muscles recoil to force air back out of the lungs. This makes way for new air, which enters when the diaphragm contacts for the next breath.

I'll use a practical example to summarize using our air-breathing hero. Let's say our hero is sitting down having a rest and maybe a frosty drink to celebrate their latest feat. While at rest, they are breathing easy, and their resting muscles can take their time producing ATP through the aerobic pathway. All of a sudden, our hero perceives trouble and gets up and begins running. In a span of a few seconds, they have expended all their stored ATP, so phosphocreatine molecules stored within their leg muscles are cleaved into creatine and phosphate to form new ATP. After around 10 seconds the phosphocreatine is also expended. This early into their exertion they have not yet started to breathe deeper or more rapidly, so their muscles have a deficit of oxygen. Glucose is broken down into pyruvate to produce ATP, and without oxygen present, pyruvate is converted to lactate. As they continue to run anaerobically, the accumulation of lactate stimulates inspiratory and expiratory muscles, and our hero begins to breathe both deeper and more frequently. This new supply of oxygen is carried through the blood to the muscles and once its supply is provided for, pyruvate can enter the Krebs cycle and undergo oxidative phosphorylation. If they maintain a constant pace, their body will be able to maintain this oxygen delivery, and they will be able to continue running for a sustained period. The maximal rate at which our hero can take in, deliver, and convert oxygen into ATP is known as their VO_2max (remember Chapter 2). However, if they have to increase their speed and the energy requirement of the task exceeds the rate their cardiorespiratory system can deliver oxygen, their muscles will begin to be deficient again in oxygen, and they will need to supplement increasingly with the anaerobic pathway. Once our hero has saved the day and returns to their seat, their body returns to primarily oxidative metabolism, phosphocreatine stores are replenished, lactate is buffered and removed from their body, and breathing returns to normal rates (at least until trouble finds our hero again).

Aquatic Respiration

Mammalian lungs are incredibly well-suited to breathing air. Animals like fish, however, spend their entire lives underwater without exposure to air. Despite the lack of air, fish still rely heavily on aerobic metabolism and thus still need oxygen.

Water, or H_2O, is comprised chemically of hydrogen and oxygen. Though this means oxygen is intrinsically a part of water, this oxygen is very strongly bound to hydrogen and cannot be easily separated, so it is not available for respiration. However, oxygen is present (in varying amounts) as a dissolved gas, and it is the dissolved oxygen that fish draw from the water. They do so not with lungs but by use of their gills (Figure 9.3). Gills are organs situated on the sides of their head/neck and are connected to the equivalent of the fish's trachea. The gills contain string-like filaments that span across the opening in arches. As water passes through the fish's mouth, it flows down the "trachea" and over the gills on its way back into the ambient water. The filaments contain dense networks of capillaries, which, like the alveoli-capillary mechanism of lungs, provide substantial surface area between the blood flowing through capillaries and the water flowing through the gills. In many aquatic species, gas exchange is also enhanced by the direction of flow: blood flows against the current of the water, allowing for more organized interaction between oxygenated water and the blood. Some species, like sharks, rely on forward movement to draw water over their gills. Other species draw water in through their mouths and with muscular contraction, they force the water through their gills. By doing so, fish can extract the small amounts of oxygen dissolved in water to sustain their energy needs, like mammals do from the air.

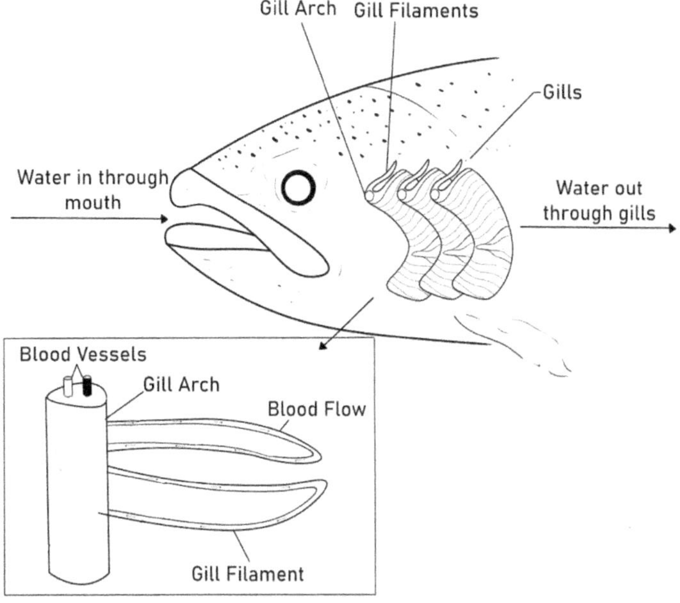

Figure 9.3: Structure and anatomy of the fish respiratory system.

Super Fish

The abilities of some animals to obtain oxygen from air and others to obtain oxygen from water is a marvellous example of the diversity of life on Earth. The ability to extract oxygen from both, however, gives an individual an incredible advantage. As with many superhuman abilities, there are precedents in nature. There are several species of fish that use their gills when oxygenated water is available but if that water becomes compromised or runs out, they have adapted to extract oxygen from the air.[92] A fish called the Desert Goby has gills and respires in water as described above. However, when oxygen levels in the water get too low or if tidal movements strand them, even without lungs they can extract oxygen from the air.[93] They do so by either drawing air into their mouth where capillaries lining the inside of their mouths extract oxygen, or through capillaries in their skin exposed to air. As a result, the Desert Goby can survive in waters with incredible ranges of salinity, temperature, pH, and oxygen levels.[93] Perhaps in the fish world Desert Gobies are thought of as superheroes for their air-breathing superpower!

Underwater Breathing

Using super-fish like the Desert Gobi as our example, it is easy to imagine the opposite ability (humans extracting oxygen from water) could be accomplished by the opposite means. That is, our hero draws water into their lungs and the oxygen from within the water diffuses into the capillaries of the alveoli. Breathing liquid using mammalian lungs can actually be done.[94] The limiting factor with water is the amount of dissolved oxygen present is generally insufficient. However, other liquids can dissolve much higher amounts of oxygen. For example, substantial amounts of oxygen can be dissolved into liquids called perfluorocarbons.[94]

In the 1960s, researchers discovered that if they allowed anesthetized cats or mice to "breathe" liquid perfluorocarbons into their lungs, gas exchange occurred as normal.[95] This has since been used in emergency medical applications by providing liquid with very high amounts of oxygen directly into patients' lungs, which they can breathe.[96] This example provides proof that mammalian lungs are capable of extracting oxygen from liquids. Alternatively, if our hero had gill-like organs that were more suited to extracting oxygen from water, they may be able to "breathe." These gills could be located within the trachea, bronchi or bronchioles, and when water is drawn into the lungs and exhaled out, the water would pass along these gills, where capillaries would draw oxygen out of the water as those in the alveoli do from air.

The major issue with both these methods is the amount of oxygen that is required. Water contains substantially less oxygen than air does. Liquid ventilation of perfluorocarbons works only if sufficient oxygen has first been dissolved in the liquid.[94] Fish are cold-blooded and move effortlessly through water, doing as little as possible to keep their overall energy requirements down. A human underwater, particularly a superhuman performing marvellous feats, would have an energy demand likely exceeding the availability of oxygen in water.

Another option is our hero could also then decrease their reliance on the aerobic pathway and shift to reliance on anaerobic metabolism. Simply, our hero would produce the ATP they require through anaerobic metabolism like certain bacteria do. Anaerobic capacity is a trainable quality: if you compare athletes trained for sprint-type activities to non-athletes there are marked differences.[97] The amount of stored pyruvate and the enzymes responsible for converting pyruvate into lactate and buffering the accumulation of hydrogen are present in much greater concentrations in an athlete's muscles. This allows them to achieve greater rates of anaerobic ATP production. Theoretically, these capacities could also be enhanced, though probably not to the point where anaerobic metabolism could be the sole mechanism.

The final potential mechanism, then, is avoiding the need for oxygen altogether. The reason electrons can combine with oxygen and drive the aerobic pathway is because oxygen has a positive charge and a high affinity for electrons. This allows it to adopt the electron into its electron cloud. However, hydrogen ions (H^+), which are produced during glycolysis, also have a positive charge and are present in abundance (although they have a lower affinity than oxygen does to electrons). If our hero's cells were able to use hydrogen in place of oxygen to receive the electron, oxygen wouldn't be required for the process at all, and our hero would be able to utilize aerobic pathways without needing to breathe oxygen. Although, they would still need to exhale carbon dioxide. Thus, they could meet their energy needs without even breathing at all, or with the minimal quantities of oxygen present in water!

Additional Considerations

Swimming and Breathing

Breathing underwater isn't overly helpful unless our hero is able to swim exceptionally well. Their energy demands may be satisfied but if they are slow, have difficulty maneuvering, or move very inefficiently, they will quickly lose their advantage. In order for underwater breathing to be an effective ability, our hero must also be an exceptional swimmer.

Conclusion

Our lungs are wonderfully suited for breathing air, much like gills are suited for water. If our hero were to be better able to extract dissolved oxygen through their existing lungs or a gill-lung hybrid organ, they could breathe oxygen (in limited amounts) from the water. Combine this with an increased ability for anaerobic metabolism, and our hero may be able to satisfy their energy needs. Best yet, if our hero's cells could use hydrogen ions as the electon terminal for aerobic metabolism instead of oxygen ions, they would be as effective, if not better, at producing ATP in water than you or I are out of water. Assuming they are an effective swimmer, this would provide them with many advantages over their villains.

CHAPTER 10

Heroes and Villains

With the exception of bionics, the abilities described in this book though theoretically "possible" would require a substantial scientific "accident," specific genetic mutations, or a moderate-to-large evolutionary step. Because of this, it is incredibly unlikely that in our lifetimes we will face villains like our heroes do (unless they arrive from outer space and their abilities are attributed to alien physiology, but that's completely another matter altogether!). However, we do face many more practical and urgent villains that we can still rely upon science to help us combat.

The storyline in *The Incredibles 2*[98] is that people have stopped believing in the ability of superheroes to protect them and they outlaw heroes from using their abilities. Similar storylines have been played out for *X-Men, Spiderman, Superman, the Avengers,* and *Batman.* For example, Batman takes the fall for the death of Harvey Dent in *The Dark Knight*[99] and is forced into hiding by the public call for his arrest. The real-life analogue of this is a loss of faith or trust in science. Many famous scientists worked against public opinion and even persecution in their development of scientific theories. Galileo was made to renounce his evidence that the Earth orbited the Sun, a truth we now unequivocally accept.[100] Similarly, Charles Darwin published *On the Origin of*

Species fourteen years after he had written it, and it is thought he did so partly because he was afraid of being persecuted for reporting his findings.[101]

As an extreme modern example, there is an organization that deems itself the "Flat Earth Society," and endeavours to promote their belief that the world is not a spherical planet but that it is flat. In 240 BC, a fellow named Eratosthenes living in Greece measured shadows from a stick in two locations at noon of the summer solstice. By comparing the differences in the length of the shadows produced by the same stick at the same time of day in different locations, he was able to calculate the circumference of the Earth almost exactly to what we know it is today, around 40,000 km.[102] The roundness of the Earth was confirmed in proof between 1505 and 1512 when Ferdinand Magellan effectively circumnavigated the globe on a ship. Thus, we have not only known the planet was round but had a remarkably good idea of its circumference since 240 BC. However, Flat Earth Society members disbelieve in images of our planets, the mathematical and physical evidence, observations of tides and the moon cycles (Earth's reflection on the moon), the ability to circumnavigate the globe on a boat or plane…we can go on all day. Ironically, they use the internet to spread their message, though the internet is only available because of satellites orbiting our spherical globe.

The fact that a tiny organization doesn't accept scientific evidence regarding the shape of our planet doesn't really harm anyone. However, a much more dangerous movement has arisen and is promoting opposition to vaccinations. Science has provided us the resources to wipe out and prevent many diseases, but they are becoming unutilized. In 1963 the measles vaccine was introduced to the United States.[103] This vaccine was further refined, and a second routine dose was implemented in 1989. As a result, measles was effectively eradicated in the United States in 2000. With the emergence of the anti-vaccination movement vaccination rates have dropped in the US, and measles has returned, with 555 cases of measles reported between January 1 and April 11, 2019.[104] Around the same time, measles cases increased 15-fold across Europe between 2016 and 2018 (just two years!).[105] Even forgoing the flu vaccine can have a substantial effect on population health. There are individuals who are unable to receive the flu vaccination because they have specific (very rare) allergies, or they react very negatively. Fortunately, large-scale population vaccination provides herd immunity where those who are unable to receive the vaccine are protected by those around them who do. However, this is only effective if enough of the population actually receives the vaccines.

Opinion, tradition, superstition, and personal anecdote don't devalue science; the only thing that can discredit science is more science. The examples of scientific illiteracy are seemingly endless. Unsupported fads are replacing medical treatment. We would rather trust celebrities than researchers (actual experts) for health information. We engage in behaviours that are harmful such as smoking and inactivity. Scientific curricula are suppressed, and science research programs continue to be defunded. Though we are unlikely to face supervillains like those depicted in comic books and movies, science still has a role in producing the heroes we need to combat the villains we do face. Science is improving treatment and care for diseases like cancer and heart disease. We have evidence-based guidelines for nutrition and exercise as preventative medicine. Vaccines protect us from preventable diseases. Space exploration has allowed for satellite technologies, and we are developing new and efficient ways to meet our increasing demand for energy. Ideas from science fiction are becoming science every day. We are in the midst of a scientific revolution that is exponentially advancing our knowledge and capabilities as fast as we can keep up. Through science, we are producing the very heroes we need for the real villains that we face. If we let science work like Batman in *The Dark Knight Rises*[106] or the Parr family in *The Incredibles 2*[98], it will help us face our villains. Who knows, maybe science will even create a superhero one day. We will never know unless we let it.

ACKNOWLEDGMENTS

I gave a presentation at the 2016 Calgary Comic Expo entitled "The Physiology of Superhero Super Strength." Considering other presenters at the expo included William Shatner, Peter Mayhew, and Stan Lee, I think my session was actually relatively well-attended. In that talk, I presented more or less an early version of Chapter 1 of this book: I described the physiology of muscle and how it could possibly be modified to lead to super strength. Following that talk, I had some engaging discussions with a colleague of mine, Dr. Ian Smith, on the merits of my ideas and, as an extension, some ideas about how super speed might be achievable. It was a few years later that I put pen to paper (or more accurately, fingers to keyboard) but that presentation in 2016, and subsequent discussions, was largely the spark for writing this book. So, the very first thank you is due to Ian for those and many other discussions we have had about muscles, family, and life.

I owe a great deal to my graduate supervisor, Dr. Walter Herzog, for my professional and academic development. I have been fortunate to learn countless lessons from him as my graduate training shaped who I am as a scientist. Further, I wrote the first draft of this book while in the final year and a half of my doctoral training. As such, I also owe a special thank you to him for allowing me the time to pursue all my interests (including writing this book) rather than writing my thesis.

Early in my scientific career, I was fortunate to train under Dr. Nicole Culos-Reed. In addition to being an excellent researcher, she is particularly passionate about translating research into sustainable community programs. Among many other things, she instilled in me the importance of communicating research beyond academic circles and that well-conducted research doesn't make a difference unless it is communicated to those who benefit or learn from it. I cited many well-conducted studies throughout this book, and I hope by reading along you learned something (or several things) new. Thank you to Nicole for these (and many other) lessons.

In my dedication, I referred to my personal heroes who have mentored me within and outside of science. If I were to try to list them all here, the list would drag on for several pages, so I won't attempt to. However, thank you to everyone on that very long list.

Several friends and professional and academic colleagues have read parts of or all of this book and provided me valuable feedback. In particular, I'd like to thank Adam, Amanda, Andrew, Jade, Randi, Seong-won, and Shannon, for reading through early drafts and providing feedback instrumental for reaching the final draft you are now holding (and hopefully enjoyed or are about to enjoy).

Thank you to my editor, Rachel, for editing out my typos, putting the punctuation in the right places, and for turning my manuscript into a book. Thank you also to Karen and Mario from Get Your Book Illustrations for turning the ideas in my head into the figures and illustrations you saw throughout the book.

Lastly, to my family: my wife is an unending source of support and constantly drives me to bigger and better things. Thank you for your constant support and for carefully combing through drafts of this book. Thank you to Xavier, who was mentioned a few times, for taking me for walks when I needed a break from the lab, data analysis, or writing. Finally, most of all, thank you to my daughters who make me feel like a superhero every day.

ABOUT THE AUTHOR

Kevin Boldt completed his academic training (BSc (Hon), MSc, PhD) at the University of Calgary in Exercise Physiology and Biomechanics. His research interests focus on understanding how the body adapts to exercise and diet, and ways to enhance these adaptations to improve health and optimize performance. His research has been well recognized with both national and international research accolades and awards. Most notably, he has received the prestigious Vanier Canada Graduate Scholarship, the Alberta Innovates Health Solutions-Clinician Fellowship, and the National Sciences and Engineering Research Council Graham Bell Award. His research has also been recognized by the International Society of Biomechanics, the Canadian Society of Biomechanics, the American Society of Biomechanics, and the Canadian Society for Exercise Physiology. He has been widely published in academic journals and has co-authored two textbook chapters. *Superhero Physiology* is his first book.

In addition to his academic work, Kevin is a Clinical Exercise Physiologist through the Canadian Society for Exercise Physiology and a Certified Strength and Conditioning Specialist though the National Strength and Conditioning Association. He uses these certifications to translate the findings of his research to clinical populations and elite athletes. He works alongside professional and Olympic athletes, firefighters, weekend warriors, and those with clinical conditions such as multiple sclerosis, stroke, arthritis, cancer, and traumatic brain injuries.

When he isn't teaching, in the laboratory, or working with clients, he spends his time reading, playing guitar, or working on his own fitness. He lives in London, ON with his wife, daughters, and dog.

Learn more at www.superherophysiology.com

REFERENCES

1. *Captain America: The First Avenger;* 2011.
2. Joanisse S, Lim C, McKendry J, Mcleod JC, Stokes T, Phillips SM. Recent advances in understanding resistance exercise training-induced skeletal muscle hypertrophy in humans. *F1000Research.* 2020;9.
3. Schoenfeld BJ. The mechanisms of muscle hypertrophy and their application to resistance training. *The Journal of Strength & Conditioning Research.* 2010;24(10):2857-2872.
4. Huxley H, Hanson J. Changes in the cross-striations of muscle during contraction and stretch and their structural interpretation. *Nature.* 1954;173:973–976.
5. Huxley A, Niedergerke R. Structural changes in muscle during contraction; interference microscopy of living muscle fibres. *Nature.* 1954;173:971–973.
6. Finer JT, Simmons RM, Spudich JA. Single myosin molecule mechanics: piconewton forces and nanometre steps. *Nature.* 1994;368:113–119.
7. Gordon AM, Huxley AF, Julian FJ. The variation in isometric tension with sarcomere length in vertebrate muscle fibres. *J Physiol.* 1966;184:170–192.
8. Herzog W. The multiple roles of titin in muscle contraction and force production. *Biophysical reviews.* 2018;10(4):1187–1199.
9. Eriksen HK, Kristiansen JR, Langangen Ø, Wehus IK. How fast could Usain Bolt have run? A dynamical study. *American Journal of Physics.* 2009;77(3):224–228.
10. Wilson AM, Lowe JC, Roskilly K, Hudson PE, Golabek KA, McNutt JW. Locomotion dynamics of hunting in wild cheetahs. *Nature.* 2013;498(7453):185–189.
11. Denny MW. Limits to running speed in dogs, horses and humans. *Journal of Experimental Biology.* 2008;211(24):3836–3849.

12. Dawson TJ, Taylor CR. Energetic cost of locomotion in kangaroos. *Nature*. 1973;246(5431):313–314.

13. Hunter JP, Marshall RN, McNair PJ. Interaction of step length and step rate during sprint running. *Medicine & Science in Sports & Exercise*. 2004;36(2):261–271.

14. Claflin DR, Faulkner JA. Shortening velocity extrapolated to zero load and unloaded shortening velocity of whole rat skeletal muscle. *The Journal of Physiology*. 1985;359:357–363.

15. Ranatunga KW. The force-velocity relation of rat fast-and slow-twitch muscles examined at different temperatures. *The Journal of Physiology*. 1984;351:517.

16. Schluter JM, Fitts RH. Shortening velocity and ATPase activity of rat skeletal muscle fibers: effects of endurance exercise training. *American Journal of Physiology-Cell Physiology*. 1994;266:C1699–C1673.

17. Rome LC, Syme DA, Hollingworth S, Lindstedt SL, Baylor SM. The whistle and the rattle: the design of sound producing muscles. *Proceedings of the National Academy of Sciences*. 1996;93:8095–8100.

18. Rome LC, Cook C, Syme DA, et al. Trading force for speed: why superfast crossbridge kinetics leads to superlow forces. *Proceedings of the National Academy of Sciences*. 1999;96:5826–5831.

19. Hudson PE, Corr SA, Wilson AM. High speed galloping in the cheetah (Acinonyx jubatus) and the racing greyhound (Canis familiaris): spatiotemporal and kinetic characteristics. *Journal of Experimental Biology*. 2012;215(14):2425–2434.

20. Hill AV. The heat of shortening and the dynamic constants of muscle. *Proceedings of the Royal Society of London B: Biological Sciences*. 1938;126:136–195.

21. Tihanyi J, Apor P, Fekete GY. Force-velocity-power characteristics and fiber composition in human knee extensor muscles. *European Journal of Applied Physiology and Occupational Physiology*. 1982;48:331–343.

22. Faria EW, Parker DL, Faria IE. The science of cycling: factors affecting performance—Part 2. *Sports Medicine*. 2005;35(4):313–338.

23. Gowans C. *Philosophy of the Buddha An Introduction*. Taylor and Francis; 2013.

24. Crombie IM. *Plato on Knowledge and Reality*. Routledge; 2013.

25. Hankinson RJ. Galen's Anatomy of the Soul. *Phronesis*. 1991;36(2):197–233.

26. Gaukroger S. *Descartes' System of Natural Philosophy*. Cambridge University Press; 2002.

27. Hodgkin AL, Huxley AF. Currents carried by sodium and potassium ions through the membrane of the giant axon of Loligo. *The Journal of Physiology.* 1952;116(4):449–472.

28. Kajiura SM, Holland KN. Electroreception in juvenile scalloped hammerhead and sandbar sharks. *Journal of Experimental Biology.* 2002;205(23):3609–3621.

29. Kalmijn AJ. The electric sense of sharks and rays. *Journal of Experimental Biology.* 1971;55(2):371–383.

30. Heffner HE, Heffner RS. Hearing ranges of laboratory animals. *Journal of the American Association for Laboratory Animal Science.* 2007;46(1):20–22.

31. Sliney DH. How light reaches the eye and its components. *International Journal of Toxicology.* 2002;21(6):501–509.

32. Walmsley R. Growth of Bone. *British Journal of Nutrition.* 1952;6(1):410–415.

33. Hales S. Vegetable Staticks: Or, an Account of Some Statical Experiments on the Sap in Vegetables: Being an Essay Towards a Natural History of Vegetation. Also, a Specimen of an Attempt to Analyze the Air, by a Great Variety of ChymioStatical Experiments; Which Were Read at Several Meetings Before the Royal Society. Vol 1. W. and J. Innys and T. Woodward; 1727.

34. Villemure I, Stokes IA. Growth plate mechanics and mechanobiology. A survey of present understanding. *Journal of Biomechanics.* 2009;42(12):1793–1803.

35. Fritzsch H. Elementary Particles: Building Blocks of Matter.; 2005.

36. Pauling L. *General Chemistry.* Dover Publications; 2014.

37. Lappert MF, Murrell JN. John Dalton, the man and his legacy: the bicentenary of his Atomic Theory. *Dalton Transactions.* 2003;(20):3811–3820.

38. Davis EA. Discovery of the electron: commentary on J. J. Thomson's classic paper of 1897. *Philosophical Magazine Letters.* 2007;87(5):293–301.

39. Thomson JJ. XL. Cathode rays. *The London, Edinburgh, and Dublin Philosophical Magazine and Journal of Science.* 1897;44(269):293–316.

40. Rutherford E. LXXIX. The scattering of α and β particles by matter and the structure of the atom. *The London, Edinburgh, and Dublin Philosophical Magazine and Journal of Science.* 1911;21(125):669–688.

41. Bohr N. The Quantum Postulate and the Recent Development of Atomic Theory. *Nature Publishing Group;* 1928.

42. Berry RS. How good is Niels Bohr's atomic model? *Contemporary Physics*. 1989;30(1):1–19.

43. Hartley HB. Antoine Laurent Lavoisier 26 August 1743—8 May 1794. *Proceedings of the Royal Society of London Series Biological Sciences*. 1947;134(876):348–377.

44. Capitaine N, Guinot B, Klioner S. Proposal for the redefinition of the astronomical unit of length through a fixed relation to the SI metre. *Rencontres de l'Observatoire Journées*. Published online 2010:20–22.

45. Rainville S, Thompson JK, Myers EG, et al. World Year of Physics: A direct test of E = mc2. *Nature*. 2005;438(7071):1096.

46. Prigogine I, Géhéniau J, Gunzig E, Nardone P. Thermodynamics of cosmological matter creation. *Proceedings Of The National Academy of Sciences*. 1988;85(20):7428–7432.

47. Berg HE, Larsson L, Tesch PA. Lower limb skeletal muscle function after 6 wk of bed rest. *Journal of Applied Physiology*. 1997;82(1):182–188.

48. Doughty D, McNichol L. Wound, Ostomy and Continence. Nurses Society Core Curriculum.; 2015.

49. Maeda K. New method of measurement of epidermal turnover in humans. *Cosmetics*. 2017;4(4):47.

50. Gantwerker EA, Hom DB. Skin: Histology and Physiology of Wound Healing. *Clinics in Plastic Surgery*. 2012;39(1):85–97.

51. KiecoltGlaser JK, Marucha PT, Mercado AM, Malarkey WB, Glaser R. Slowing of wound healing by psychological stress. *The Lancet*. 1995;346(8984):1194–1196.

52. Silverstein P. Smoking and wound healing. *The American Journal of Medicine*. 1992;93(1):S22–S24.

53. Hoffman M, Harger A, Lenkowski A, Hedner U, Roberts HR, Monroe DM. Cutaneous wound healing is impaired in hemophilia B. *Blood*. 2006;108(9):3053–3060.

54. Sela J, Itai B. Principles of Bone Regeneration.; 2012.

55. Lieberman J, Friedlaender G. Bone Regeneration and Repair Biology and Clinical Applications.; 2005.

56. Dinsmore CE. A History of Regeneration Research: Milestones in the Evolution of a Science. Cambridge University Press; 1991.

57. Monaghan JR, Maden M. Cellular plasticity during vertebrate appendage regeneration. In: *New Perspectives in Regeneration*. Springer; 2012:53–74.

58. Seifert AW, Monaghan JR, Voss SR, Maden M. Skin regeneration in adult axolotls: a blueprint for scar-free healing in vertebrates. *PloS one*. 2012;7(4).

59. Tanaka EM. The molecular and cellular choreography of appendage regeneration. *Cell*. 2016;165(7):1598–1608.

60. Hu MS, Borrelli MR, Hong WX, et al. Embryonic skin development and repair. *Organogenesis*. 2018;14(1):46–63.

61. Ferguson MW, O'Kane S. Scar–free healing: from embryonic mechanisms to adult therapeutic intervention. *Philosophical Transactions of the Royal Society of London Series B: Biological Sciences*. 2004;359(1445):839–850.

62. Nodder S, Martin P. Wound healing in embryos: a review. *Anatomy and Embryology*. 1997;195(3):215–228.

63. Jevotovsky DS, Alfonso AR, Einhorn TA, Chiu ES. Osteoarthritis and stem cell therapy in humans: a systematic review. *Osteoarthritis and Cartilage*. 2018;26(6):711–729.

64. Morton RW, Murphy KT, McKellar SR, et al. A systematic review, meta-analysis and meta-regression of the effect of protein supplementation on resistance training-induced gains in muscle mass and strength in healthy adults. *Br J Sports Med*. 2018;52:376-384.

65. Nerlich AG, Zink A, Szeimies U, Hagedorn HG. Ancient Egyptian prosthesis of the big toe. *The Lancet*. 2000;356(9248):2176–2179.

66. Markatos K, Karamanou M, Saranteas T, Mavrogenis AF. Hallmarks of amputation surgery. *International Orthopaedics*. 2019;43(2):493–499.

67. Camporesi S. Oscar Pistorius, enhancement and posthumans. *Journal of Medical Ethics*. 2008;34(9):639–639.

68. Müller R, Tronicke L, Abel R, Lechler K. Prosthetic pushoff power in transtibial amputee level ground walking: A systematic review. *PloS one*. 2019;14(11).

69. Pursley RJ. Harness patterns for upper-extremity prostheses. *Artificial Limbs*. 1955;2(3):26.

70. Popov B. The bioelectrically controlled prosthesis. *The Journal of Bone and Joint Surgery British Volume*. 1965;47(3):421–424.

71. Sahyouni R, Mahmoodi A, Chen JW, et al. Interfacing with the nervous system: A review of current bioelectric technologies. *Neurosurgical Review*. 2019;42(2):227–241.

72. Ajiboye AB, Willett FR, Young DR, et al. Restoration of reaching and grasping movements through brain-controlled muscle stimulation in a

person with tetraplegia: a proof-of-concept demonstration. *The Lancet.* 2017;389(10081):1821–1830.

73. Hamzaid NA, Yusof NHM, Jasni F. Sensory Systems in MicroProcessor Controlled Prosthetic Leg: A Review. *IEEE Sensors Journal.* Published online 2019.

74. Bogue R. Exoskeletons and robotic prosthetics: a review of recent developments. *Industrial Robot: an International Journal.* Published online 2009.

75. Mooney LM, Rouse EJ, Herr HM. Autonomous exoskeleton reduces metabolic cost of walking. In: *2014 36th Annual International Conference of the IEEE Engineering in Medicine and Biology Society.* IEEE; 2014:3065–3068.

76. Mooney LM, Herr HM. Biomechanical walking mechanisms underlying the metabolic reduction caused by an autonomous exoskeleton. *Journal of Neuroengineering and Rehabilitation.* 2016;13(1):4.

77. Leclair J, Pardoel S, Helal A, Doumit M. Development of an unpowered ankle exoskeleton for walking assist. *Disability and Rehabilitation: Assistive Technology.* 2020;15(1):1–13.

78. Zeilig G, Weingarden H, Zwecker M, Dudkiewicz I, Bloch A, Esquenazi A. Safety and tolerance of the ReWalk™ exoskeleton suit for ambulation by people with complete spinal cord injury: a pilot study. *The journal of Spinal Cord Medicine.* 2012;35(2):96–101.

79. Young AJ, Ferris DP. State of the art and future directions for lower limb robotic exoskeletons. *IEEE Transactions on Neural Systems and Rehabilitation Engineering.* 2016;25(2):171182.

80. Shi D, Zhang W, Zhang W, Ding X. A Review on Lower Limb Rehabilitation Exoskeleton Robots. *Chinese Journal of Mechanical Engineering.* 2019;32(1):74.

81. Kazerooni H, Racine JL, Huang L, Steger R. On the control of the berkeley lower extremity exoskeleton (BLEEX). In: *Proceedings of the 2005 IEEE International Conference on Robotics and Automation.* IEEE; 2005:4353–4360.

82. Dollar AM, Herr H. Lower extremity exoskeletons and active orthoses: Challenges and stateoftheart. *IEEE Transactions on Robotics.* 2008;24(1):144–158.

83. Alabdulkarim S, Nussbaum MA. Influences of different exoskeleton designs and tool mass on physical demands and performance in a simulated overhead drilling task. *Applied Ergonomics.* 2019;74:55–66.

84. Takahashi N, Takahashi H, Koike H. A Novel Soft Exoskeleton Glove for Motor Skill Acquisition Similar to Anatomical Structure of Forearm Muscles. In: *2019 IEEE Conference on Virtual Reality and 3D User Interfaces (VR)*. 2019:1568–1569. doi:10.1109/VR.2019.8797919

85. *Alien*; 1979.

86. *Avatar*; 2009.

87. *Iron Man*; 2008.

88. Henry CJK. Basal metabolic rate studies in humans: measurement and development of new equations. *Public Health Nutrition*. 2005;8(7a):1133–1152.

89. Cartaxana P, Trampe E, Kühl M, Cruz S. Kleptoplast photosynthesis is nutritionally relevant in the sea slug Elysia viridis. *Scientific Reports*. 2017;7(1):1–10.

90. Thauer RK, Jungermann K, Decker K. Energy conservation in chemotrophic anaerobic bacteria. *Bacteriological Reviews*. 1977;41(1):100.

91. Vazzana I, Rizzo M, Dara S, Niutta PP, Giudice E, Piccione G. Haematological changes following reining trials in quarter horses. *Acta Scientiae Veterinariae*. 2014;42(1):1–5.

92. Nilsson GE, Hobbs JA, ÖstlundNilsson S, Munday PL. Hypoxia tolerance and air-breathing ability correlate with habitat preference in coral-dwelling fishes. *Coral Reefs*. 2007;26(2):241–248.

93. Thompson GG, Withers PC. Aerial and aquatic respiration of the Australian desert goby, Chlamydogobius eremius. *Comparative Biochemistry and Physiology Part A: Molecular & Integrative Physiology*. 2002;131(4):871–879.

94. Kohlhauer M, Berdeaux A, Kerber RE, Micheau P, Ghaleh B, Tissier R. Liquid ventilation for the induction of ultrafast hypothermia in resuscitation sciences: a review. *Therapeutic Hypothermia and Temperature Management*. 2016;6(2):63–70.

95. Clark LC, Gollan F. Survival of mammals breathing organic liquids equilibrated with oxygen at atmospheric pressure. *Science*. 1966;152(3730):1755–1756.

96. Wolfson MR, Shaffer TH. Pulmonary applications of perfluorochemical liquids: ventilation and beyond. *Paediatric Respiratory Reviews*. 2005;6(2):117–127.

97. Degens H, Stasiulis A, Skurvydas A, Statkeviciene B, Venckunas T. Physiological comparison between non-athletes, endurance,

power and team athletes. *European Journal of Applied Physiology.* 2019;119(6):1377–1386.

98. *Incredibles 2*; 2018.

99. *The Dark Knight*; 2008.

100. Hawking S. On the Shoulders of Giants: The Great Works of Physics and Astronomy.; 2002.

101. Van Wyhe J. Mind the gap: Did Darwin avoid publishing his theory for many years? *Notes and records of the Royal Society.* 2007;61(2):177–205.

102. Longhorn M, Hughes S. Modern replication of Eratosthenes' measurement of the circumference of Earth. *Physics Education.* 2015;50(2):175.

103. Phadke VK, Bednarczyk RA, Salmon DA, Omer SB. Association between vaccine refusal and vaccine-preventable diseases in the United States: a review of measles and pertussis. *JAMA.* 2016;315(11):1149–1158.

104. Paules CI, Marston HD, Fauci AS. Measles in 2019—going backward. *New England Journal of Medicine.* 2019;380(23):2185–2187.

105. Thornton J. *Measles Cases in Europe Tripled from 2017 to 2018.* British Medical Journal Publishing Group; 2019.

106. *The Dark Knight Rises*; 2012.

www.ingramcontent.com/pod-product-compliance
Lightning Source LLC
Chambersburg PA
CBHW050947050726

47592CB00007B/2464